Frontiers in Aging Science

Volume 1

Challenging Aging
The Anti-senescence Effects of Hormesis, Environmental Enrichment and Information Exposure

Authored By

Marios Kyriazis
ELPIS Foundation for Indefinite Lifespans
London,
United Kingdom

advertisements or ideas contained in the Work.

Limitation of Liability:

In no event will Bentham Science Publishers, its staff, editors and/or authors, be liable for any damages, including, without limitation, special, incidental and/or consequential damages and/or damages for lost data and/or profits arising out of (whether directly or indirectly) the use or inability to use the Work. The entire liability of Bentham Science Publishers shall be limited to the amount actually paid by you for the Work.

General:

1. Any dispute or claim arising out of or in connection with this License Agreement or the Work (including non-contractual disputes or claims) will be governed by and construed in accordance with the laws of the U.A.E. as applied in the Emirate of Dubai. Each party agrees that the courts of the Emirate of Dubai shall have exclusive jurisdiction to settle any dispute or claim arising out of or in connection with this License Agreement or the Work (including non-contractual disputes or claims).
2. Your rights under this License Agreement will automatically terminate without notice and without the need for a court order if at any point you breach any terms of this License Agreement. In no event will any delay or failure by Bentham Science Publishers in enforcing your compliance with this License Agreement constitute a waiver of any of its rights.
3. You acknowledge that you have read this License Agreement, and agree to be bound by its terms and conditions. To the extent that any other terms and conditions presented on any website of Bentham Science Publishers conflict with, or are inconsistent with, the terms and conditions set out in this License Agreement, you acknowledge that the terms and conditions set out in this License Agreement shall prevail.

Bentham Science Publishers Ltd.
Executive Suite Y - 2
PO Box 7917, Saif Zone
Sharjah, U.A.E.
Email: subscriptions@benthamscience.org

BENTHAM SCIENCE

CONTENTS

Biography

Marios Kyriazis qualified as a medical doctor (MD) from the University of Rome, Italy, and after preclinical work in the USA he worked as a clinician in acute medicine in Cyprus, and the UK. He subsequently qualified as a Gerontologist with interest in the biology of ageing and became a Chartered Member of the academic organisation 'Royal Society of Biology' in the UK. He also has a post-graduate qualification in Geriatric Medicine from the Royal College of Physicians of London. Other appointments include Member of the Board of Trustees at the Mediterranean Graduate School of Applied Social Cognition, affiliate researcher at the Evolution, Complexity and Cognition Group, University of Brussels, and a Ronin Research Scholar.

Currently, he works with the ELPIs Foundation for Indefinite Lifespans, a serious endeavour to study the elimination of age-related degeneration. The research is focused on transdisciplinary models and explores common principles between biology, complexity sciences, evolution, cybernetics, neurosciences, and techno-cultural elements. Areas of interest include robustness and degeneracy in organic systems, fragility and redundancy, repair processes (including self-repair), hormesis and environmental enrichment in ageing, and prolonged survival of somatic cells.

Dr. Kyriazis is a member of several editorial boards including the Elsevier Editorial System, Rejuvenation Research, The Biologist, World Journal of Translational Medicine, Peptides journal, the European Journal of Clinical Nutrition etc. He is also a Member of many age-related organisations, committees and advisory boards. He has a portfolio of over 1000 articles, papers and lectures in the field of healthy ageing.

FOREWORD

Life is a constant struggle between the intrinsic and extrinsic challenges and the ability to counteract, defend and adapt to those challenges. This is a highly dynamic process. However, beyond the evolutionarily required "essential lifespan" (ELS), a progressive failure of homeodynamics is the fate of life, manifested in ageing and eventual death. How to prevent or slow down this failure, or how to enhance our functionality and survival, is the challenge of ageing.

Comprehensive scientific studies over the past five decades have led to the conclusion that the traditional disease-orientated biomedical model of ageing needs to be abandoned for a wholistic concept. This is because there are no specific gerontogenes or any master controller molecule(s) that cause ageing and which can be counteracted one at a time. Maintaining health, and any possible extension of lifespan, require whole-body interventions. One such approach is the phenomenon of mild stress-induced hormesis, which is the subject matter of this book.

Hormesis, hormetins and hormetics are, respectively, the phenomenon, the agents and the science of bi-phasic dose response to stressors. This is a scientific and evidence-based approach. Marios Kyriazis' book, which also includes a contribution from another author, deals with this subject of scientific and wholistic hormesis in a most comprehensive and accessible manner.

Hormesis works. Stress of choice is good for health. Food, physical activity and psycho-social engagement are the three pillars of health. These can be personalised and optimised by using the three principles of hormesis: pleasure, moderation and variety. This book helps in opening several new practical possibilities in this regard.

Dr. Suresh Rattan
International Association of Gerontology and Geriatrics European Region – (IAGG-ER)
Aarhus University
Aarhus
Denmark

PREFACE

This Series on 'Frontiers in Aging Science' aims to exploit the method of online sharing of information, and will examine in detail several important models, hypotheses, theories and other notions (including Blue Skies Research) which may help us elucidate the intricacies of the ageing process. The first such eBook will examine the role of challenges. These are interventions that provoke action (a protective response) from the organism. This response is mediated by the up-regulation of protective cellular mechanisms that diminish the effect of age-related degeneration in humans.

Several authors have suggested that the process of biological ageing is associated with loss of information, disruption of homeostasis, and a reduction of functional and physiological complexity. As time-related damage accumulates, and the processes of repair become progressively less able to deal with this damage, organisms begin to experience dysfunction, degeneration and chronic clinical diseases, eventually leading to death. One possible way of remediating this loss of complexity is to increase exposure to relevant and meaningful information which can, through various mechanisms, up-regulate functional and structural complexity with a consequent improvement in function. This basic premise (that age-related loss of complexity may be counteracted by exposure to stimulation and information) has been studied in a variety of levels and under many different guises. One way of increasing information exposure is through mild and repeated challenges or mild stress, *i.e.* hormesis. In medicine and biology hormesis is defined as 'an adaptive response of cells and organisms to a moderate, intermittent, challenge'.

Hormesis describes phenomena where there is a low dose stimulation, high dose inhibition, and it suggests that nutritional, physical, mental and chemical challenges, if appropriately timed, may result in mild damage to the organism which up-regulates repair mechanisms. In crude terms it can be said that during the process of repairing this damage, any coincidental age-related damage is also repaired. A similar concept is that of Environmental Enrichment where experimental animals are exposed to an enriched environment with regards to visual, auditory and habitat augmentation. The majority of experiments confirm that an enriched and stimulating environment (an 'information-rich' habitat) has several positive effects on health, specifically on brain and immune function. These concepts are presented in Chapters 1-3.

In Chapters 4 and 5 there is a discussion about the biological mechanisms of information exposure, and how the impact of new information and challenges may result in the reallocation of resources from the germ line to the soma. This is important because it may underlie a hitherto dormant mechanism that may lead to radical reduction of age-related degeneration. In Chapter 6 I analyse further the relationships between information, challenges, human evolution and possible biological changes, building upon the previous discussion. In Chapter 7, Atanu Chatterjee from the Indian Institute of Technology analyses certain significant concepts which are crucial in our understanding of life generally and the human body in particular. He provides a grounding and a framework for expanding our notions of hormesis and stimulation, from a general domain to the specific case of human ageing. He explains notions such as complexity in living systems, energy distribution and entropy, which form the foundation of our efforts to devise ways that maximise information exposure that increases our biological complexity.

Finally, in chapter 8 I expand the scope of the discussion and aim to examine ways we can become better participants within an increasingly technological environment. The overall aim is to examine ways whereby humans, in a modern context, can harness the power of challenging information, and use it to up-regulate their functional complexity (both in the biological and in the social sense). As a result, damage repair becomes maximised, the risk of dysfunction diminishes and the incidence and prevalence of age-related degeneration and disease is kept to a minimum or even totally eliminated.

Our methodology is conceptually different from many existing approaches which depend on physical, pharmacological, genetic or cellular methods and other disruptive technologies for defying ageing. Despite discussing several drugs or compounds acting as hormetic agents, the essential characteristic of our methodology is that it is based less on physical items and more on environmental abstract, virtual and cognitive elements (see Figure below).

The title of the book reflects this: Ageing is seen as a challenging problem but, at the same time, it may be overcome by exposure to challenges (situations that incite biological action) which may be clinically useful. The book is targeted at gerontologists, anti-ageing physicians and clinicians, medical students, and university students (in gerontology, biology, complexity sciences, evolution). The slant is a blend of graduate/postgraduate level, providing an in-depth analysis and discussion of the main concepts (hormesis, environment, information, human evolution, biology of ageing) and a synthesis of these as applied to defying age degeneration. It will be valuable to those pursuing a medical or biological career, as well as others interested in human ageing.

Master Figure. This explains the relationship, influences and feedback loops between the concepts of hormesis, environmental enrichment and social effects upon somatic and germ line biology. The intention is to show that external challenges have an impact on somatic repair mechanisms and thus may help improve repair of age-related degeneration, resulting in prolongation of healthy lifespan. The reader is requested to revisit this figure after reading the entire book and review the interconnections of the different components. This will give a much clearer idea of the overall concept I tried to describe in this book.

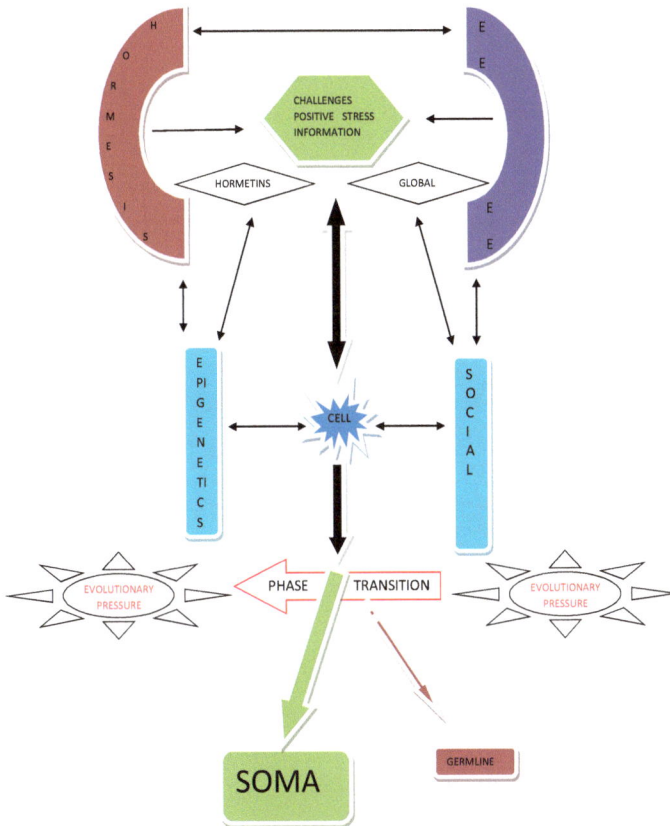

Fig. (1). Schematic representation of the relationship between the different concepts discussed in this book. The concept of hormesis is based on challenges, positive stress and information exposure. It influences (partly through Hormetins) both directly and indirectly (through epigenetic mechanisms) the cell. (The concept of the 'cell' is better described as a 'somatic agent' which includes cells, molecules, genetic material and anything else that makes a human, with the exception of germline material). Environmental enrichment (EE) is also based on challenges, positive stress and information, and through both local and global mechanisms (such as the Global Brain, smart cities, Ambient Intelligence) also has an effect on the 'cell'. This effect is modulated by social and cultural factors. The enormous evolutionary pressure placed upon the 'cell' results in a phase transition which shifts the priority of repair resource allocation from the germline to the soma, resulting in reduced or absent age-related functional decline.

CONFLICT OF INTEREST

The author confirms that author has no conflict of interest to declare for this publication.

ACKNOWLEDGEMENTS

I would like to thank the contributors and reviewers of this project for their input, discussions and inspiration. Suresh Rattan for acting as a compass in the entire field of biogerontology generally, and for his contributions to the hormesis research specifically. Ed Calabrese and Mark Mattson, for providing the sources of a substantial amount of research mentioned in the book. Francis Heylighen and the members of the Evolution Complexity and Cognition group at the University of Brussels, for horizon-expanding discussions, arguments and inspiration in developing concepts such as the Global Brain, Complex adaptive system behaviour, evolution and cybernetics.

Dr. Marios Kyriazis
ELPIS Foundation for Indefinite Lifespans
London
United Kingdom

Frontiers in Aging Science

Volume 1

Challenging Aging
The Anti-senescence Effects of Hormesis, Environmental Enrichment and Information Exposure

2

CHAPTER 1

Hormesis and Adaptation

Abstract: Our biological response to external challenges frequently obeys hormetic principles. During the phenomenon of hormesis, mild stressful challenges may up-regulate defence and repair pathways, with a subsequent overall improvement in function. It is important to highlight that hormesis is a dose-response, non-linear phenomenon, meaning that a low dose of a stressor can result in benefit whereas a higher dose may result in damage. Hormesis is invoked when the challenge is of sufficient magnitude and appropriate quality as to satisfy the definition of 'novelty'. Routine and monotony do not, as a rule, invoke a hormetic response. In this chapter I will discuss certain characteristics of hormesis as applied to humans, and examine several situations whereby an adequately-timed stimulus may be of practical health benefit. The assessment and response to the new challenge leads to adaptation and thus, eventually, improvement of function within a particular environment (the environment where the challenges have originated from). In this way, there is a direct link between external challenging information and internal physical or biological changes. This link will be explored in detail, both in this chapter and in other chapters of this book.

Keywords: Adaptation, Cellular networks, Exploratory behaviour, Homeodynamic space, Hormesis, Non-linearity, Novelty, Physical challenges, Power law, Stress response, Stressor, Stimulation.

SOME DEFINITIONS

Agent = An entity that acts on its environment.

Stress = Any sudden, unforeseen perturbation of a system, when the system itself does not have the resources to deal with the change, *i.e.* it cannot, or has not have the time to, adapt to the perturbation.

Stressful event = A perturbation of a system, a challenge, a stimulus that incites the system to act.

Marios Kyriazis

Challenge = The term refers to a softer situation where, following a stressful event the system has both the time and the capability to adapt, *i.e.* change in response to the perturbation. So, strictly, stress and challenge are not equivalent, but mild/positive stress (as opposed to chronic, intense stress) can be seen as the equivalent of a 'Challenge'. Here, a challenge is defined as a situation that potentially carries biological value for an organism, so that the organism is inclined to act. A challenge provokes action because it represents a situation in which not acting will lead to an overall lower fitness than acting.

Adaptation = When the system or the agent undergoes a structural or functional rearrangement in order to accommodate the new information carried by the challenging event.

Hormesis = A biphasic dose response to an environmental challenge, characterized by a low dose stimulation (benefit) and a high dose inhibition (damage).

Indefinite and Infinite Lifespans An **indefinite** lifespan is a lifespan without a pre-determined end, and it denotes the virtual elimination of the mortality rate as a function of age (the elimination of age-related functional decline). In other words, the incidence of involuntary death caused by ageing tends to zero. Death can still ensue through other means such as accidents, injuries, infections, starvation and so on. An **infinite** lifespan on the other hand, is equivalent to true immortality and the total abolition of death from any cause (however idealistic this may be).

Ageing (in humans) = Time-related dysfunction.

Evolution = The adaptation to changes in the environment, so that survival continues.

Fitness = Good function within a specific environment.

INTRODUCTION

Although hormesis is a term applicable to a wide range of situations [1] in this book I discuss hormesis with particular relevance to humans. Hormesis is a phenomenon characterised by a non-linear, 'U'-shaped, 'low-dose activation, high-dose inhibition' principle. In other words, a low dose of a stimulus can positively challenge the organism and result in health benefits, whereas an excessive, suboptimal, or prolonged exposure can result in damage and disease [2]. In a wider sense, the concept is based on repeated mild exposure to new information, a sustained (but not excessive) state of 'novelty' which resets homoeostatic mechanisms.

In order to use the correct terminology, it is necessary to clarify that there are different terms describing diverse hormetic effects. For example, a previous exposure to mild stress (a low dose of a hormetic agent) protects the organism against a larger, stressful dose later on. This is called 'Conditioning Hormesis' and it bestows a protective effect against future stresses [3]. A different type of hormesis is 'Post-exposure Conditioning Hormesis', when an organism who is subjected to a high, toxic level of stress may experience improvements when it is subsequently exposed again to low doses of the same stressor. For the purposes of this book, while we use the term 'hormesis' we, on the whole, refer in fact to the Conditioning Hormesis aspect of the concept.

The hormetic response is triggered by encounters with any physical, chemical, biological, mental or other challenges, which may disturb the cellular or organismic homeostatic mechanisms [4] (Fig. **1**). Although hormetic effects have been studied extensively at cellular level, and not so much at the level of the entire organism, it is legitimate to assume that effects at higher levels are real and valid [5].

Hormesis, through functionally diverse stressors, may trigger several mechanisms of lifespan extension such as disruption of the Insulin-like Growth Factor-1 (IGF-1) signalling pathway, up-regulation of immunity, proteostasis and oxidative stress response [7]. A basic characteristic of a hormetic event is **novelty of information**. Here, novelty is defined as 'the quality of being new, original, or unusual', and this includes both unfamiliarity and unconventionality. Novelty is also associated with creativity, innovation and imagination. These are essential characteristics because, as I will discuss in other chapters, a worldview which encompasses creativity, innovation and imagination is more likely to lead to a healthy and prolonged lifespan, exactly because it invokes novelty, which results to positive hormetic change.

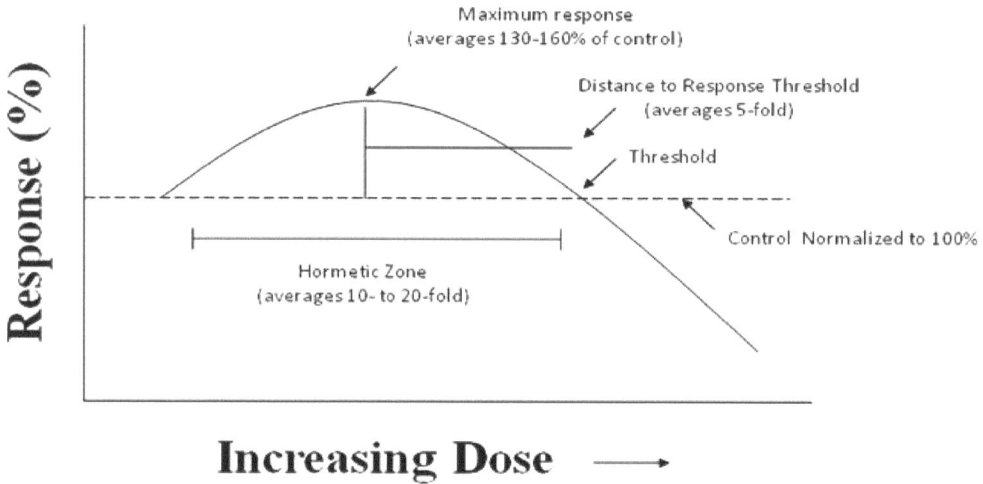

Fig. (1). The hormesis (biphasic) dose–response curve: generalized quantitative features. The hormetic dose response is characterized by specific quantitative features concerning its amplitude and width. Based on more than 10,000 hormetic dose responses in various hormetic databases, approximately 80% have their maximum response less than twofold greater than the control. The strong majority of hormetic dose responses extend over approximately a 5- to 10-fold dose range immediately below the estimated response threshold.... The hormetic response appears to quantitatively describe the limits of biological plasticity across phyla as well as at different levels of biological organization (cell, organ, and organism). Image and text credit from [5].

Hormetic Stimulation

When we are exposed to a new stimulus, (be that a physical, chemical, mental, or other), our sensory systems detect the new information (in the form of nerve impulses, or otherwise) which is carried from the periphery to central processing areas. There, the information is assessed and used in order to initiate processes that lead to changes in the configurations of existing networks, including genetic, epigenetic or any other networks (Fig. **2**) [8]. Therefore, there is a direct link between information and physical changes or modulation of our biology. This link will be explored in detail in several parts of this book.

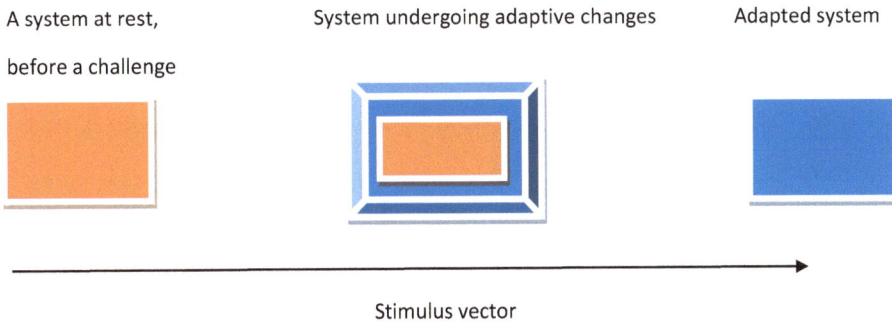

A system at rest,   System undergoing adaptive changes   Adapted system

before a challenge

Stimulus vector

Fig. (2). System adaptation following stimulation. A system at rest is being challenged by a novel stimulus, which causes the entire system to undergo a process of internal changes, aiming to buffer the stimulus. Eventually the system attains a different state compared to the original. This system is now 'better' than the original in the sense that it can respond to further similar stimuli without exhibiting dysfunction, and is thus well-suited to its new environment (which contains that stimulus).

Scott [9] has commented that humans have managed to survive for millions of years while being continually exposed to ionising radiation (both from cosmic and solar origin). In common with other mammals, this has helped us develop a continually-evolving protection system, what he calls an 'activated natural protection' (ANP) system which is regulated by epigenetic factors, *i.e.* factors which respond to environmental challenges. An optimal activation of ANP results in hormetic phenotypes which exhibit increased resistance to stress. As a result, these phenotypes are healthier compared to phenotypes which are poorly resistant to stress. This is relevant in ageing where adaptation to external and internal stresses must remain effective. One way effective adaptation can be achieved is *via* epigenetic mechanisms. Hormesis causes epigenetic adaptation, and an optimal balance between over or under activation of genes [10]. In this respect, hormesis can reverse the dysregulation in epigenetic control which is seen in ageing. The role of epigenetic and environmental influences in age-related dysfunction will be examined in detail in other chapters.

It has been argued that the stress-induced benefits of hormesis may not come free but may, in fact, come with trade-offs particularly with regards to immunity (*i.e.* hormesis may result in reduced immune function) [11]. This is at odds with other studies which show that hormesis improves immune status, as well as other processes, and it has been critiqued [12]. In their critique, Le Bourg and Rattan conclude that:

> *It seems that the balance between positive and possible negative effects of mild stress is clearly on the positive side, and any trade-offs in fitness are specific to the general health, robustness and resilience of the body.*

In addition, it is also well accepted that exposure to chronic stress (*i.e.* not mildly challenging, but intense and sustained stress) can indeed result in immunosuppression [13], as well as other deleterious events such as increased cortisol, and disruptions of homoeostasis. In population studies, when an organism is exposed to a mild challenge, there is a period of sustained growth and the death rates are dampened. When the stressful stimulation exceeds a certain threshold, then death rates increase and growth rates decrease [14]. The organism becomes less sensitive to the stressor if it has already been exposed at low levels of that stressor [6]. Therefore, this general phenomenon suggests that previous exposure to mild challenges and information that compel the organism to act, fortifies the organism against future severe stresses and it is thus beneficial to the health of that organism.

Of course, there are inter-individual differences, making the response to a challenging stimulus more variable. In addition, genetic and environmental factors are relevant, and we may reach a situation where many different factors act synergistically or antagonistically to modify the expected result. Nevertheless, as a general concept, hormesis provides a useful framework for practical suggestions that can be applied in everyday life [15, 16].

The Homeodynamic Space

One example of the general principle where a stimulus can be toxic at some doses and beneficial at others is the case of reactive oxygen species (ROS). Agents such as hydrogen peroxide, superoxide and others may prevent age-related degeneration by acting as hormetic agents or as signalling molecules [17]. What matters here is the dose: a low exposure to these agents can improve systemic defences and regulate adaptive processes. Based on this premise, it may be argued that meticulous avoidance of oxidative events such as artificial blocking of ROS signalling through oral supplementation with antioxidants, may prevent this hormetic signalling and result in damage [18].

The concept of homeodynamic space proposed by the biogerontologist Suresh Rattan is useful when considering hormesis, and it can help in understanding the relationship between low or excessive stimulation [19]. He quotes:

> *Aging, senescence and death are the final manifestations of unsuccessful homeostasis or failure of homeodynamics. A wide range of molecular, cellular and physiological pathways of repair are well known, and these include multiple pathways of nuclear and mitochondrial DNA repair, free radical counteracting mechanisms, protein turnover and repair, detoxification mechanisms, and other processes including immune- and stress-responses. All these processes involve numerous genes whose products and their interactions give rise to a "homeodynamic space" or the "buffering capacity", which is the ultimate determinant of an individual's chance and ability to survive and maintain a healthy state. A progressive shrinking of the homeodynamic space, mainly due the accumulation of molecular damage, is the hallmark of aging and the cause of origin of age-related diseases.*

It is interesting here to encounter the concept of 'buffering' which was also further developed by cyberneticist Francis Heylighen [20]. He quotes:

> *Aging is analyzed as the spontaneous loss of adaptivity and increase in fragility that characterizes dynamic systems. Cybernetics defines the general regulatory mechanisms that a system can use to prevent or repair the damage produced by disturbances. According to the law of requisite variety, disturbances can be held in check by maximizing*

buffering capacity, range of compensatory actions, and knowledge about which action to apply to which disturbance. This suggests a general strategy for rejuvenating the organism by increasing its capabilities of adaptation. Buffering can be optimized by providing sufficient rest together with plenty of nutrients: amino acids, antioxidants, methyl donors, vitamins, minerals, etc. Knowledge and the range of action can be extended by subjecting the organism to an as large as possible variety of challenges. These challenges are ideally brief so as not to deplete resources and produce irreversible damage. However, they should be sufficiently intense and unpredictable to induce an overshoot in the mobilization of resources for damage repair, and to stimulate the organism to build stronger capabilities for tackling future challenges. This allows them to override the trade-offs and limitations that evolution has built into the organism's repair processes in order to conserve potentially scarce resources. Such acute, "hormetic" stressors strengthen the organism in part via the "order from noise" mechanism that destroys dysfunctional structures by subjecting them to strong, random variations. They include heat and cold, physical exertion, exposure, stretching, vibration, fasting, food toxins, micro-organisms, environmental enrichment and psychological challenges. The proposed buffering-challenging strategy may be able to extend life indefinitely, by forcing a periodic rebuilding and extension of capabilities, while using the Internet as an endless source of new knowledge about how to deal with disturbances....... The reason for the overshoot is the uncertainty about the seriousness of the challenge: better hit an attacker hard enough the first time so that he won't come back, than run the risk of a second attack that may kill you. If the potential destructiveness of the challenge is not known, the safer strategy is to mobilize more than may be needed. On the other hand, for well-known challenges, the organism can estimate exactly how much it will need, and save the rest. The effect of the overshoot is that more damage will be repaired than the one caused by the disturbance, thus actually rejuvenating the organism. Moreover, the organism will have learned to expect more intense challenges than it was used to, and therefore to build up its capabilities for tackling similar problems in the future. This increases its range of action or "homeodynamic space".

Therefore, it is becoming clear that there is an interplay between challenges and buffering mechanisms, which aims to avoid extremes of stimulation and to maintain the system within a defined space of function [21]. Any extreme fluctuations may result in deviations outside the homeodynamic space, and loss of function. It is thus important to try and maintain the quality and strength of the challenging stimulation within certain limits. Examples of such hormetic challenges (hormetins) applicable to everyday life and their mechanisms have been discussed by Rattan [22, 23] and I will discuss some examples in the next chapter.

The three main characteristics of homeodynamic space are:

a. Stress Response
b. Damage Control, and
c. Continuous Remodelling [24]

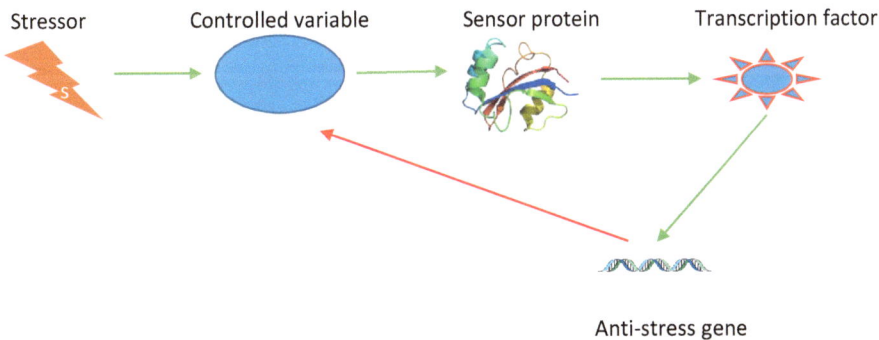

Anti-stress gene

Fig. (3). The negative feedback loop involved in the regulation of the stress response. This helps maintain the function of the different genes and proteins within the limits of the homeodynamic space. Physical, nutritional, cognitive and other stressors act on a host of variables such as reactive oxygen species, DNA adducts and glucose, which activate sensor proteins that regulate transcription factors. These then activate anti-stress genes which, through negative control, normalise the controlled variable, and thus the system remains buffered within a certain normal landscape. [Figure adapted from [25]].

The concept of 'Continuous Remodelling' is important as it supports the notion that our internal biological processes respond in a way not only to counteract the external stressful stimulus, but also to improve local and systemic structure and function (Fig. **3**).

Cellular Networks during Hormetic Stress

Stress (including mild stress) can affect the function and configuration of cellular networks [25]. In higher organisms, cells are connected in a dynamical network fashion, where there is continuous rearrangement of connections, depending on the degree of stimulation (external or internal) [26]. The links between cells can be weak ('weakly-connected' cells) or strong ('strongly-connected' cells, usually forming hubs) (Fig. **4**). Szalay *et al.* [28] showed that chronic stress which is perceived as a threat can decrease the density of links between cells and results in diminished function. Strong stress may also induce a phase transition where the cell completely changes its way of operating.

Cells usually form 'small words' (when the elements of any network are separated by only a small number of other elements [29] - (see Fig. **4**). This is useful when it comes to communication and cross-talking between cells. Obviously, it would be ideal if a signal can propagate from one cell to another without the need to pass through many and repeated paths involving many cells. Strongly-connected cells form hubs which facilitate communication and are also more resilient to damage. Following a stressful stimulus, the network may exhibit a 'Le Chatelier-type network principle' where after the stimulation, the network reduces the strength of the links, and this makes it more difficult for the perturbation to propagate [28]. When the perturbation is exhausted, the links remain in the existing configuration, and, as a result, the network has now achieved a new configuration, *i.e.* it has adapted as in Fig. (**2**). The network has thus become better able to withstand a further perturbation.

Examples of Hormetic Challenges

As explained above, it is possible to experience hormetic effects through exposure to novelty and through exploratory behaviour [30]. The combination of novelty and unpredictability eventually leads to increased creativity, which is a unique human characteristic. And, of course, as it is discussed in this book, may also help improve health and longevity: it is known that hormetic challenges may be used in order to reduce the impact of ageing [31]. Exploratory, goal-oriented behaviour, in areas where one is unfamiliar is a legitimate source of hormetic challenges [32]. The novelty of information exposure triggers several biological and behavioural mechanisms which perturb homoeostasis and initiate a series of actions which, on the whole, result in an improved state compared to the state before the action: 'improved', in the sense that the organism is now better adapted to its surroundings. This exploratory behaviour may include physical elements (such as exploring new routes, new objects or new items), or cognitive elements (new concepts, new reading material, new ideas). Of course, a combination of these two elements is ideal for an all-round effect.

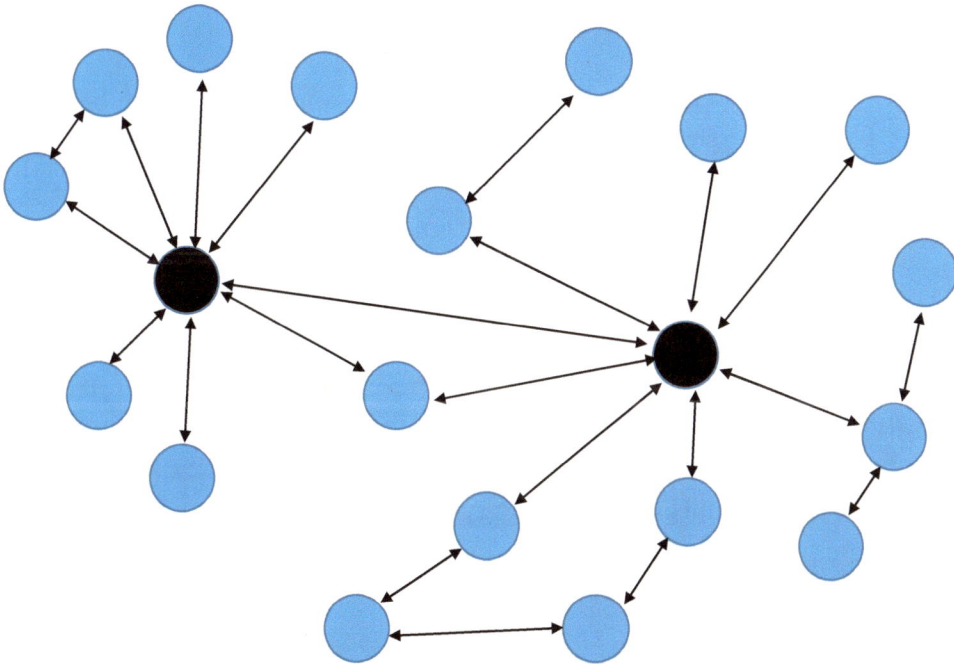

Fig. (4). 'Small-world', scale-free cellular networks. Schematic representation of transmission of information between individual cells which are connected with other cells. Most cells in this network (blue) interact with only one or two other cells (weakly-connected cells) whereas two (black) are connected with many other cells simultaneously (strongly-connected cells, forming hubs). In this example, these strongly-connected cells can communicate directly with 75% of other cells, whereas weakly-connected cells interact with just 10% of others. Random damage (or repair of damage) in some weakly-connected cells is not likely to cause significant disruption (or benefit) to the network, whereas damage to, or repair of strongly-connected cells is going to affect huge parts of the network (from [27]). Random attrition of cells in the network eventually leads to network collapse, and underlies degenerative disease.

A 'Bias for action' is another behavioural way for achieving stimulation and challenging our cognitive processes [33]. Exhibiting a bias for action refers to active decision-making, a general propensity to act 'on the spot' without having the full information regarding the task. It is 'making ideas happen', *i.e.* acting purposefully. It does not mean however, to act without forethought. It refers to having novel ideas based on previous knowledge and acting fast on those ideas in order to make them reality. The decisive action involves a certain degree of risk without being excessively risky, a typical, dose-dependent, hormetic concept. Humans (and some animals) frequently engage in behaviour which has no clear adaptive basis, but it can be seen as 'intrinsically motivated, and often directed toward the unknown and the unexpected' [34].

Many practical ways of inducing other mild hormetic stress in animals have been described [35]. For instance, it is known that heat stress in cows induces changes in reproduction [36]. Research in humans is less detailed but there are indications that we can relatively safely extrapolate at least some results from animal studies to humans. Therefore, when it comes to practical hormetic advice, we may consider issues such as:

1. **Avoid excessive and ritual cleanliness, and allow some exposure to dirt**. It is important to allow micro-organisms to interact with the immune system in order to maintain the effectiveness of the immune function. The continual exposure to pathogens up-regulates elements of the immunity repair system which can then be useful when a more pronounced accidental exposure takes place. Several studies have shown a relationship between pathogen exposure and subsequent immunity response. For a summary see [37]. Dirt, pathogens, minor injuries and minor infections act as hormetic challenges to the immune system, up-regulating appropriate stress responses and allowing the immune system to respond more effectively in a future stressful encounter.

2. Another challenge is based on the same principle, and involves aiming for (or, at least, tolerating) **activities which may result in trivial/mild injury** [38]. For

example, gardening without hand protection may result in minor grazes or scratches which act as challenging stimuli to the immune system, and incite it to become more effective at fighting infections. As a consequence, this will help heal more serious wounds or injuries at a later stage (this may well include age-related damage).

The general concept of hormetic protection against future stressful events has been studied in a variety of experiments involving insect bites [39, 40]. It was shown that bee stings (and exposure to bee venom) can induce immune changes such as T-cell proliferation and improved IgE response, which protect against a future toxic event. This is an example of Conditioning Hormesis [41]. It was also suggested that while high exposure to mosquito bites has a negative impact on immunity, a lower but sustained exposure has immune system benefits [42].

3. **Direct exposure to wind, sunshine or rain** may also be considered as mid hormetic challenges. Air pressure and other sensors on the skin (such as thermal sensors [43] for example), register the sensory information carried by the wind or rain, and transmit the information to the brain which, in turn, modulates compounds that can repair skin damage, restore any changes in temperature, and heal skin irritation. Through this action, the neuro-immune mechanisms remain active and are ready to respond to a subsequent more serious challenge. This principle is exemplified by research in possums. Researchers have shown that exposure to mild wind and light rain improves the immune system of laboratory possums, and increases their ability to cope in a much colder or wetter environment in the future [44].

Not forgetting, of course, that lack of exposure to sunlight may result in vitamin D deficiency. At the initial stages this causes muscle aches and bone pains, tiredness and lack of energy. Vitamin D also supports the immune system and it is therefore essential during ageing. Thus, exposure to the elements should bear a hormetic basis: some exposure is good, too much is detrimental (and it may result in sunburn dehydration, premature skin ageing and similar).

Other studies have supported the general principle of heat or cold exposure as a means for regulating subsequent immune response, at least in animals [45]

(Box **1**).

BOX 1. Sauna as a heat stress: Health benefits of sauna used as a hormetic challenge

- Following heat stress, the expression of HSP(Heat Shock Proteins) increase and this mediates hormetic health benefits [46] as discussed above. This can be translated into some relatively well-studied clinical treatment modalities [47]. For example, it has been suggested that sauna at around 80-100 degrees Celsius for 15 minutes three times a week for three months can have a positive impact on fasting glycaemia, glycated haemoglobin, body weight, and body fat content [48]. This may be mediated through nitric oxide and HSP70 expression.
- In a study, six healthy volunteers were exposed to hyperthermia treatment (20 min/day, 40°C, 100% relative humidity, using a nano-mist sauna method). The concentration of cortisol, catecholamines and leukocyte activity were studied. It was found that cortisol and adrenaline levels decreased, while B and T cell function improved [49].
- A prospective cohort study of 2315 middle-aged (age range 42-60 years) men from Eastern Finland showed that increased frequency of sauna bathing can result in a reduced risk of sudden cardiac death, fatal coronary heart disease, fatal cardiovascular disease, and all-cause mortality [50].
- Another study of Finnish sauna on healthy men (three 15 minute sessions every other day) showed a statistically significant reduction of LDL, total cholesterol and triglycerides. There was also a small increase in HDL levels. These results were comparable to those experienced through moderate exercise [51].
- Benefits following hyperthermia (sauna) treatments were also reported by patients with chronic obstructive airway disease, chronic fatigue, persistent pain, or addiction [52]. It is considered to be a safe treatment modality, with no significant side effects apart from its use in pregnancy which may be teratogenic.

These and other studies show that the hormetic effects of heat challenge (*via* sauna for example) work through a variety of mechanisms to improve overall health and reduce the risk of mortality.

4. **A 'power-law' lifestyle**. This acts as a continual and changeable challenge to several organs and systems, up-regulating overall function. Some details regarding this type of lifestyle are given in the section below.

Power Law Lifestyle

Ageing is a universal phenomenon and so it must obey universal laws and principles. Certain mathematical principles that apply globally can thus safely be applied in ageing. One of these laws concerns the function of 'power law' and fractals. Power law is a mathematical relationship between two quantities. In the case of hormesis, consider the relationship that exists between the **intensity** (i) of any activity as a power of its **frequency** (f): $f^{(i)}$.

In practice, it has been suggested that in order to follow a more 'natural' and therefore evolutionarily successful lifestyle, the frequency of any given activity (such as exercise, for instance) needs to vary as a power of its intensity: frequent low-intensity exercises, less frequent medium-intensity exercises, and rare high-intensity ones [53].

A good example of power law lifestyle is encountered in the Paleolithic concept. The modern notion of the Paleolithic lifestyle refers to a lifestyle that exposes us to situations similar to those experienced by humans living in the wild, natural environment over 10 000 years ago. The notion refers to both the nutritional [54] and physical challenges [55] (such as Paleo diet and the Paleo lifestyle), and is based upon the concepts of hormesis and power law. The Paleo diet suggests consuming foods which were eaten by ancient hunter-gatherers such as nuts, raw vegetables, meat and fish, as well as avoiding foods from agricultural origins such as cereals, rice, sugar, and other processed food or foods with high glycaemic index [56].

Within the Paleo model, it is necessary to mimic natural patterns of eating, such as skipping meals or eating at irregular intervals, and follow Intermittent Fasting methodologies. These introduce elements of novelty, unpredictability and positive challenges, all of which modulate the stress response.

The Paleo diet umbrella also encompasses two important nutritional concepts, that of calorie restriction and that of intermittent fasting. Calorie restriction is as a form of nutritional hormesis [57]. Intermittent Fasting is based on the 'feast or famine' pattern normally encountered in the wild. This aims to achieve a positive nutritional stress response [58]. The practice places the organism under nutritional stress [59], which can be beneficial if it is not prolonged. Benefits include an improvement in blood sugar and cholesterol levels, protection against neurotoxicity, enhancing the immune system and many others. An important component of the Paleo lifestyle is physical exercise. With hormesis in mind, it is possible to recommend physical activity that has power-law characteristics, meaning that it is necessary to follow an exercise regime including many low-level exercises (brisk walking, gentle swimming) and avoiding frequent intense activities such as running marathons. These physical exercises need to be performed at irregular intervals and ever-changing durations (*i.e.* avoiding repetitions) for the full hormetic effect to become apparent [60]. Power-law activities improve the sense of balance, strengthen the bones against osteoporosis, improve immunity, and exercise diverse groups of muscles. It is important to highlight once again that within the context of a Paleo lifestyle, any repeated and protracted high intensity activities (such as prolonged jogging or intense weight lifting) are not advised. These actions increase production of uncontrolled reactive oxygen species which accelerate damage to the tissues [61].

The power-law methodology favours irregular and unusual exercises, with a slant towards using natural (as opposed to artificial, man-made items) such as lifting logs or dragging trees, walking barefoot, rock and tree climbing, throwing, running for short distances in shallow water, *etc.*, all preferably performed in a natural environment such as a forest or a park. This aligns well with the concept of biophilia- a tendency to lean towards green, natural environments [62]. As mentioned above, the irregularity of the exercises is also important, as it introduces an element of novelty and continual challenge. Repetitive and extreme exercises have no adaptive value and are associated with many adverse effects [63].

NOTE: For the sake of completeness, it is worth mentioning that there is an opposite school of thought, which offers an alternative view based on evolution and modern techno-cultural society environments. This view places more emphasis on subject matters virtually **contrary** to those discussed above, *i.e.* with much less emphasis on 'healthy' diet and exercise, and considerably more emphasis on cognitive stimulation, which is more appropriate for some sections of humanity (at least in some developed countries). In this context, the meaning of the term 'healthy' or 'natural' is being questioned outside of conventional thinking. These concepts are discussed later on in the book.

Apart from nutrition and exercise, there are many other natural or artificially-inspired activities that may result in hormetic health actions (Table **1**).

Table 1. Other practical hormetic activities in humans could be.

Activity	Hormetic Action
Low dose ionising radiation (LDIR)	Exposure to mild (hormetic) radiation has been shown to modulate apoptosis and prevent untimely death of healthy cells, and at the same time promote elimination of cancerous cells [64]. This is a promising but impractical hormetic stimulation, awaiting clarification. It is known that LDIR can cause different changes to human embryonic stem cells, depending on the actual dose used [65]. It may be claimed that living within a modern-day 'radiation-smog' has hitherto poorly defined hormetic benefits.
Exposure to very low temperatures: WBC (Whole Body Cryotherapy)	This is known to improve blood circulation, stimulates immunity and has positive effects in several chronic degenerative conditions such as arthritis and hypertension. Its basis is a hormetic response, a brief exposure to a damaging agent (in this cases extreme cold) that initiates the stress response and prepares for a more serious or sustained danger [66]. Specifically, exposure to short bursts of cold stress has been proven to up-regulate the action of macrophages (*i.e.* improves immunity), whereas a more prolonged exposure diminishes their function. This is a typical example of hormesis when a short-lived challenge is beneficial, but becomes detrimental if it is prolonged [67]. It is interesting to note that H_2S (hydrogen sulphide - the hormetic poison mentioned in Box 2) can also lower the body temperature, achieving results similar to those seen with WBC
Extreme fairground rides [68]	This is a more unusual or unconventional way of experiencing a challenging stimulation. Exposure to hyper-gravity is a well-known method for achieving hormesis in the laboratory [69] and it is mentioned here to give an indication of the range of possible hormetic stimulants [70]. Experiments using centrifuges for small animals [71] or even human centrifuges [72] have shown a variety of health benefits.
The use of hormetic metabolic poisons.	These include H_2S, Cyanide, Rotenone, Antimycin and Malonate [73]. Low dose aflatoxin ingestion has been reported to benefit chickens [74]. However, as research in this area is clearly lacking, there are no meaningful practical suggestions that can be given in this respect. It is worth remembering that hydrogen sulphide is emerging as a useful metabolic and signalling molecule in low doses (see Box **2**)

(Table 1) contd.....

Activity	Hormetic Action
Short extreme heat, steam room mimicking heat shock stress	Effects similar to sauna, as discussed above

---

BOX 2. The effects of Hydrogen Sulphide

Hydrogen sulphide (H_2S) is considered here as it is a typical hormetic compound. It is poisonous in high doses, but in low dose it has a variety of physiological effects. Together with nitric oxide (NO) and carbon monoxide (CO) it is an important gaso-transmitter, a gas which participates in the transmission of signals in a biological context. Hydrogen sulphide exposure:

- affects apoptosis [75]
- has antioxidant properties [76]
- reduces inflammatory markers and improves angiogenesis [77]. The use of H_2S donors has been studied in laboratory animals and was found to be promising [78].

Sexuality: an Unmentionable Hormetic Challenge

Back in 2010, I published a paper which discussed, among other techniques, the role of sexuality in ageing, and the hormetic effects of sexual activities [79]. It is known that hormesis plays an active role in modulating the ageing of the sexual system. For instance, a dose-response phenomenon exists when low amounts of zinc in the diet can increase male fecundity patterns, whereas increasing the dose of zinc beyond an optimal level actually decreases fecundity [80]. Although this study has been performed in flies, it suggests that, based upon the universal nature of hormesis, this effect may well be true for humans too. In another study of flies it was shown that early exposure to a stressful stimulant may result in improved conditioning in later life and increased fecundity [81]. Serum testosterone levels exhibit sensitivity to hormetic stimuli such as low dose ionising radiation [82], and low doses of the poison searelonone may lead to increased secretion of oestrogens based on a hormetic effect [83]. I further quote from my paper:

"Zhang et al. [84] *have demonstrated that, although high-dose radiation caused extensive reproductive adverse changes in the mouse reproductive system, pre-exposure to low-dose radiation was beneficial. In particular, they found that pre-treatment with low-dose radiation attenuates the negative effects of high-dose radiation at a later stage and improves pituitary gonadotropins, follicular stimulating hormone (FSH), luteinizing hormone, testosterone, testicular weight, and sperm count. These are typical hormetic responses, and they support the suggestion that the reproductive and endocrine systems obey universal hormetic laws. Although more research is clearly needed in this area, it could be claimed that challenges and stimuli that cause mild physical or mental stress upon the sexual and reproductive systems could prove to be beneficial at the clinical level. One may hypothesize for example, that recommending unconventional sexual activities in consenting older people may, through hormesis, reduce the effects of age-related sexual dysfunction. These activities may act upon serotoninergic pathways and/or stimulate endorphins, BDNF, or other elements, including testosterone and estrogen concentrations* [85, 86]. *A consequence of any increased sexual activity is that it improves blood circulation and muscle strength, which acts beneficially on other organs through biological amplification.*

It has been shown that visual and tactile stimulation can have a positive effect upon patients with erectile dysfunction [87]. *Furthermore, visual erotic and vibrotactile stimulation can increase average penile circumference to a sufficient size suitable to achieve vaginal intercourse, even in impotent men up to the age of 90 years* [88]. *The argument here is that increased visual erotic inputs act as a mild hormetic challenge on the brain, which responds beneficially via neurohormonal pathways.*

On the basis of these arguments, and as long as there are no major physical barriers, physicians or other suitably qualified health practitioners could recommend a range of practical measures to enhance sexuality in the older patient. These recommendations need to go beyond simple romanticism and affection and should relate to attaining pure sexual gratification for the hormetic stimulus to become effective. It is possible to use both conventional and unconventional methods to encourage variety, stimulation, and sexual arousal (either with a partner or in solitary situations)" (Box **3**).

BOX 3. Examples of hormetically-inspired erotic activities are:

- Vibro-tactile stimulation (genital touching, masturbation, use of vibrator devices, orgasmic meditation) [89]
- Visual erotic (legal pornography, 'top shelf' magazines, 'selfies', videos, internet) [90]
- Sexual games ('sexting', heavy petting, cross dressing, sexual fantasies, mock Sadomasochism) [91]
- Unusual sexual positions (either during intercourse or otherwise, for instance consensual face-sitting) [92]
- Use of aids (unusual condoms, G-spot stimulators, touch (finger-worn) vibrators, rings, 'love balls') [93]

Therefore it can be seen that a variety of stimuli can result in beneficial health effects if used at the correct dose. I have already discussed how the frequency of applying the hormetic challenge needs to be regulated. In the following section I will discuss some aspects of evaluating the strength and amplitude of the simulation.

Regulating the Regulator: What Constitutes a 'Good' Positive Challenge

A positive challenge is a condition that requires action from our physiological processes, because it represents an opportunity to be exploited [94]. A challenge is essentially a problem that needs resolving, and our physiology is forced to take a specific action which must be for the benefit of the organism. It is forced to **select** the best option among a number of others. An appropriate selection itself creates information which reduces uncertainty [95], and it results in improvement of physiological function. With respect to the brain, any meaningful information (knowledge, experience, wisdom, excellence), *via* expressive activation of appropriate brain mechanisms (sensory to cortex, and other areas such as the prefrontal cortex and the anterior insula) activates (*i.e.* increases the energy available to, or the potential energy of) biological patterns. An improved physiological response may include better repair and maintenance of the tissues, which, in clinical terms, translates to prolongation of health-span and avoidance of age-related degeneration.

In this respect, challenge (accumulation of useful information) can prevent regress, *i.e.* reduce the rate of entropy increase, as discussed in Chapter 7. According to Shannon [96], entropy increase is associated with loss of information, entailing loss of organisation. Therefore, more (appropriate) information should equal an improvement in organisation, and thus improvement in function.

However not all information is useful, and not all information is beneficial, so it is necessary to filter it in order to avoid information overload. Information of suitable magnitude is necessary in order to improve problem-solving by our biological and physiological processes. This may be achieved if the biological process can make the appropriate selection when confronted with a challenge or a stressful stimulus. The increased power of selection means that the best choices will be selected for the ultimate benefit of the organism. The benefits of a challenge are derived not only from external information but also from internally-created abstract thoughts, meditation, awe *etc*. Intentional cognitive enhancement should be distinguished from a mere passive cognitive stimulation [97] (see below).

How to measure the degree of stress and how to ensure that the exposure is appropriate? This is a question that is asked over and over, with researchers concerned that overexposure to a stimulus may cause adverse effects, whereas underexposure may prove useless in achieving a benefit. The level of optimal stimulation can be difficult to predict. It will have to, nevertheless, be based on hormetic principles, *i.e.* low dose stimulation, high dose inhibition. The issue regarding the achievement of an optimal level of stimulation which can result in health benefits must be addressed.

One way to choose whether a stimulus is of sufficient magnitude or not, is to refer to the notion of *'challenge versus threat'* [98]. When an organism is presented with a sudden or unexpected problem, it will evaluate whether it has the resources or the ability to deal with this problem effectively. If yes, then the problem is seen as a (hormetic) 'challenge'. If the organism does not have access to sufficient resources or a range of suitable options in order to confront the problem, then this is perceived as a 'threat', *i.e.* damaging, to be avoided [99].

A very crude subjective measure of choosing if a particular stimulus is seen as a challenge or threat is to rely on an internal 'feel good' factor: if we feel comfortable with dealing with a problem then it is likely that this will exert challenging hormetic benefits on our systems. If we feel uncomfortable or distressed by a stimulus then it is likely that our physiology perceives this as a threat. This does not necessarily mean a threat to our survival but it could mean a threat to our well-being, or more precisely, a threat to our homoeostasis. There could be hormonal or neuroendocrine elements involved in this process but here I am concentrating on neural elements only. The neuro-anatomical basis of this evaluation has been placed in the anterior insula region of the brain. The anterior insula reciprocally communicates with visceral and autonomic nervous elements (including gut microbiota neuro-elements [100] and thus monitors the state of peripheral resources [101]. Activation of the anterior insula results in the subjective perception of 'effort' [102].

The results of the continual monitoring of energy resources needed to perform a task can thus be manifested as 'comfort' or 'discomfort' [103]. In addition, there may be biological markers for sensing over or under stimulation. Studies of cellular stress response show that there exist stress transducers and, for example, in the endoplasmic reticulum there are the IRE1 (inositol-requiring enzyme 1), PERK (Protein kinase RNA-like endoplasmic reticulum kinase), and ATF6 (activating-transcription-factor-6) pathways which sense any perturbations connected to stress [104, 105]. These subsequently trigger downstream signalling pathways which help deal with the stress in question. Another biomarker is the Heat Shock Protein family which are elevated following stress [106].

These transducers are able to sense not only the presence of stress but also its intensity, and subsequently induce a dose-dependent response. The presence of stress response markers provides a possible way (in principle) to monitor the intensity of any stressful stimulus, by evaluating the activity of suitably identified biomarkers following a stressful hormetic event. In general terms, it is necessary to aim for short but intense challenges followed by a certain period of recovery, in order to avoid excessive exposure and exhaustion of resources. Chronic (continual) exposure to intense cognitive stimuli is counterproductive.

I need to highlight that certain random and unstructured (*i.e.* background noise, chaotic) challenges are also a necessary element in achieving robustness of adaptability to stress. The addition of random noise into the system enhances antifragility [107]. These random, background unstructured challenges help single out and destroy any brittle, fragile configurations of our stress-response pathways, and encourage the system to explore new configurations which may carry more robustness. When these robust configurations are identified, are **selectively** retained by the system and thus making it increasingly more resistant to future perturbations. This is the phenomenon of 'order out of noise', better explained by the notion of *simulated annealing* [108].

In the search for identifying the characteristics of an optimal challenge, the 'Yerkes–Dodson law' becomes relevant. This describes the relationship between arousal and performance, and it posits that mental or physical arousal (stimulation, novelty, challenges) leads to improved performance but only when arousal is up to a certain level. If arousal is excessive or below the optimal level, then performance worsens. Too high levels of stimulation result in decreased performance, and this is based on a typical hormetic principle graphically illustrated as a 'bell'- shaped curve [109] (Fig. **5**).

It is also possible to study another way of measuring the magnitude of the challenging stimulus, by considering the concept of *flow*. This has been described by Csikszentmihalyi back in 1992 [111]. Essentially, the concept describes how a challenge that matches one's skills and abilities causes well-being. If the challenge is over one's ability then it causes anxiety. If it is below, it causes boredom. He quotes [112]:

The best moments usually occur when a person's body or mind is stretched to its limits in a voluntary effort to accomplish something difficult and worthwhile. Optimal experience is thus something we make happen.

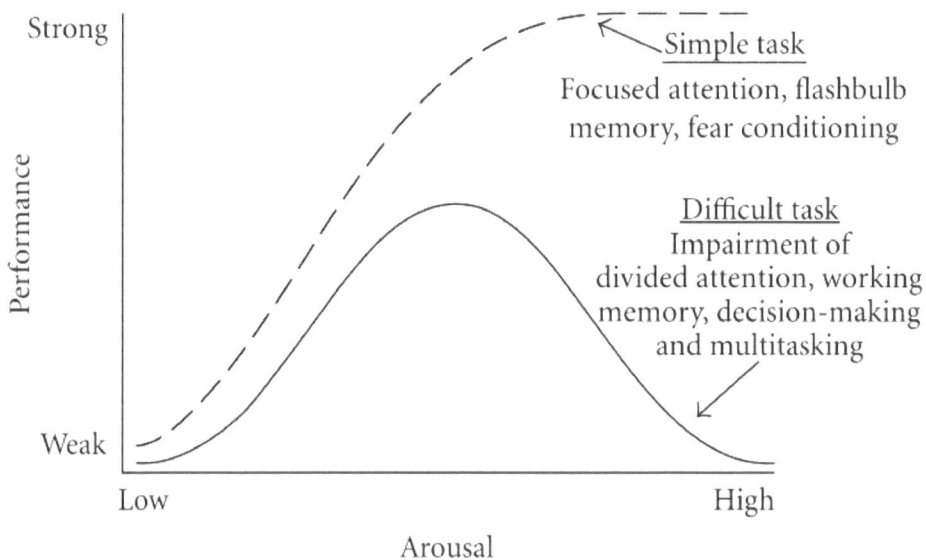

Fig. (5). The relationship between performance and arousal. [From [110]]. Here the physiological flow of information could be: exposure to a challenge causes hormonal and neuro-endocrine changes (such as rising cortisol). This causes a physiological response in the brain which may involve changes in protein or neurotransmitter function, leading to plasticity potentiation, and successful adaptation to the original challenging event.

This is similar to Blascovich's ideas of 'challenge *versus* threat' mentioned above [113]. So, if a mental challenge causes excessive stress or anxiety, is unlikely to be beneficial in ageing. If it is of such a low intensity that causes boredom, then it will not be beneficial either [114]. As mentioned, it is known that the feeling of 'effort' necessary to achieve a cognitive function depends on reactive control systems involving the anterior insula, which monitor information about energetic costs [115].

As hinted above, health benefits may be experienced not only through externally-obtained information but also through internally-originating challenges such as meditation and abstract thinking. Mindfulness meditation is a combination of training regimes which involve increased attention and heightened information flow, which act as a challenge to several brain structures. For instance, it has been shown that mindfulness mediation increases blood flow to the anterior insula, indicating that the appropriate sensors which monitor 'effort' in the brain are activated [116]. During meditation, the functional connectivity of the amygdala is also improved [117]. The amygdala is involved in the regulation and integration of emotional stressors, *i.e.* stressors that initiate internally to the body [118]. Abstract thoughts also act on the amygdala, which encodes any abstract cognitive information and modulates behaviour [119], as discussed in Chapter 3, with examples given in the Appendix. Therefore, it can be argued that internally-derived challenging and intentional behaviour acts directly on certain defined brain regions. These then integrate the signals, modulate the stress response and improve neural function, behaviour and, through the hypothalamic-pituitar--adrenal axis, also general health.

This variation between a high and a low degree of challenges can mobilize necessary resources and improve repair processes which can then strengthen functional and structural redundancy and resilience. However, the stage of information fatigue or cognitive exhaustion should not be reached. In this situation it is important to emphasize that there is no **constan**t optimal level of challenges. A challenge that can be beneficial for a person one day may prove neutral for the same person on another day (*intra*-individual variation). And a challenge which is beneficial for one person, may not be beneficial for another similar person matched for age, sex, health and other parameters (*inter*-individual variation)- see Chapter 2. In order to mimic natural patterns we need to be exposed to novel challenges which exhibit a *power law* distribution: many slight ones, some average ones, a few big ones, and very occasional huge ones.

CONCLUSION

In this chapter I attempted to introduce an element of complexity in the subject of ageing, and I used terms from network science, complexity theories and cybernetics. I accept that I sometimes use a language that is not easily recognised by biologists, but I believe this is a language (the language of complexity) that biologists need to embrace in order to deal with such a complex matter as ageing. Cybernetic terminology is used in order to explore certain elements of the biology of ageing which are difficult to understand by using a conventional biological approach [20]. For instance, in the cybernetic sense, intelligence is the ability to make consistently correct selections from available choices. This means that one has to be in a position which contains (is forced to contain) challenges that need resolving, and choices that need to be made. Routine, monotony and regularity do not account for increased need to select, whereas variability, irregularity and uncertainty maximise our need to select (and thus increase hormetic responses and information content). Physical challenges have a host of hormetically-based effects which basically aim at making the organism more adaptable to the challenging environment. The hormetic action helps up-regulate multilevel elements of our biology and maintains it in a state ready to address more serious or life threatening conditions, including age-related degeneration [120]. As a consequence, several organic parameters become adapted and maintained within the healthy landscape which is the Homeodynamic Space. In the following chapter I will discuss specific examples of natural and artificial compounds which may help achieve such hormetic adaptation.

REFERENCES

[1]   Calabrese EJ. Hormesis: from mainstream to therapy. J Cell Commun Signal 2014; 8(4): 289-91.
        [http://dx.doi.org/10.1007/s12079-014-0255-5] [PMID: 25366126]

[2]   Rattan SI, Le Bourg E, Eds. Hormesis in health and disease. Boca Raton: CRC Press 2014.
        [http://dx.doi.org/10.1201/b17042]

[3]   Wiegant FA, Prins HA, Van Wijk R. Postconditioning hormesis put in perspective: an overview of experimental and clinical studies. Dose Response 2011; 9(2): 209-24.
        [http://dx.doi.org/10.2203/dose-response.10-004.Wiegant] [PMID: 21731537]

[4]   Pardon M-C. Hormesis is applicable as a pro-healthy aging intervention in mammals and human beings. Dose Response 2009; 8(1): 22-7.
        [http://dx.doi.org/10.2203/dose-response.09-020.Pardon] [PMID: 20221284]

[5]   Cornelius C, Koverech G, Crupi R, *et al.* Osteoporosis and alzheimer pathology: Role of cellular stress response and hormetic redox signaling in aging and bone remodeling. Front Pharmacol 2014; 5: 120.
[http://dx.doi.org/10.3389/fphar.2014.00120] [PMID: 24959146]

[6]   Agutter PS. Elucidating the mechanism(s) of hormesis at the cellular level: the universal cell response. Am J Pharmacol Toxicol 2008; 3(1): 100-10.
[http://dx.doi.org/10.3844/ajptsp.2008.100.110]

[7]   Shore DE, Ruvkun G. A cytoprotective perspective on longevity regulation. Trends Cell Biol 2013; 23(9): 409-20.
[http://dx.doi.org/10.1016/j.tcb.2013.04.007] [PMID: 23726168]

[8]   Yang Y, Yeo CK. Conceptual network model from sensory neurons to astrocytes of the human nervous system. IEEE Trans Biomed Eng 2015; 62(7): 1843-52.
[http://dx.doi.org/10.1109/TBME.2015.2405549] [PMID: 25706505]

[9]   Scott BR. Radiation-hormesis phenotypes, the related mechanisms and implications for disease prevention and therapy. J Cell Commun Signal 2014; 8(4): 341-52.
[http://dx.doi.org/10.1007/s12079-014-0250-x] [PMID: 25324149]

[10]   Vaiserman AM. Hormesis, adaptive epigenetic reorganization, and implications for human health and longevity. Dose Response 2010; 8(1): 16-21.
[http://dx.doi.org/10.2203/dose-response.09-014.Vaiserman] [PMID: 20221294]

[11]   McClure CD, Zhong W, Hunt VL, Chapman FM, Hill FV, Priest NK. Hormesis results in trade-offs with immunity. Evolution 2014; 68(8): 2225-33.
[PMID: 24862588]

[12]   Le Bourg É, Rattan SI. Hormesis and trade-offs: a comment. Dose Response 2014; 12(4): 522-4.
[http://dx.doi.org/10.2203/dose-response.14-054.LeBourg] [PMID: 25552954]

[13]   Meng D, Hu Y, Xiao C, Wei T, Zou Q, Wang M. Chronic heat stress inhibits immune responses to H5N1 vaccination through regulating $CD4^+$ $CD25^+$ $Foxp3^+$ Tregs. Biomed Res Int 2013.

[14]   Rushton JP. Toward a theory of human multiple birthing: sociobiology and r/K reproductive strategies. Acta Genet Med Gemellol (Roma) 1987; 36(3): 289-96.
[PMID: 3330387]

[15]   Rattan SI. Molecular gerontology: from homeodynamics to hormesis. Curr Pharm Des 2014; 20(18): 3036-9.
[http://dx.doi.org/10.2174/13816128113196660708] [PMID: 24079765]

[16]   Kyriazis M. Nonlinear stimulation and hormesis in human aging: practical examples and action mechanisms. Rejuvenation Res 2010; 13(4): 445-52.
[http://dx.doi.org/10.1089/rej.2009.0996] [PMID: 20662589]

[17]   Ristow M, Schmeisser K. Mitohormesis: Promoting Health and Lifespan by Increased Levels of Reactive Oxygen Species (ROS). Dose Response 2014; 12(2): 288-341.
[http://dx.doi.org/10.2203/dose-response.13-035.Ristow] [PMID: 24910588]

[18]   Rahal A, Kumar A, Singh V, *et al.* Oxidative stress, prooxidants, and antioxidants: The Interplay. BioMed Res Int 2014.

[19]   Rattan SI. Aging is not a disease: implications for intervention. Aging Dis 2014; 5(3): 196-202.
[http://dx.doi.org/10.14336/AD.2014.0500196] [PMID: 24900942]

[20]   Heylighen F. Cybernetic principles of aging and rejuvenation: the buffering- challenging strategy for life extension. Curr Aging Sci 2014; 7(1): 60-75.
[http://dx.doi.org/10.2174/1874609807666140521095925] [PMID: 24852018]

[21]   Rattan SI. Biogerontology: from here to where? The lord cohen medal lecture-2011. Biogerontology 2012; 13(1): 83-91.
[http://dx.doi.org/10.1007/s10522-011-9354-3] [PMID: 21866365]

[22]   Rattan SI. Molecular gerontology: from homeodynamics to hormesis. Curr Pharm Des 2014; 20(18): 3036-9.
[http://dx.doi.org/10.2174/13816128113196660708] [PMID: 24079765]

[23]   Demirovic D, Rattan SI. Establishing cellular stress response profiles as biomarkers of homeodynamics, health and hormesis. Exp Gerontol 2013; 48(1): 94-8.
[http://dx.doi.org/10.1016/j.exger.2012.02.005] [PMID: 22525591]

[24]   Rattan SI, Demirovic D. Hormesis can and does work in humans. Dose Response 2009; 8(1): 58-63.
[http://dx.doi.org/10.2203/dose-response.09-041.Rattan] [PMID: 20221290]

[25]   Zhang Q, Pi J, Woods CG, Jarabek AM, Clewell HJ, Anderson ME. Hormesis and adaptive cellular control systems. Dose-Response. Int J 2008; 6(2): 5.

[26]   Barabási AL, Oltvai ZN. Network biology: understanding the cell's functional organization. Nat Rev Genet 2004; 5(2): 101-13.
[http://dx.doi.org/10.1038/nrg1272] [PMID: 14735121]

[27]   Kyriazis M, Apostolides A. The fallacy of the longevity elixir: Negligible senescence may be achieved, but not by using something physical. Curr Aging Sci 2015; 8(3): 227-34.
[http://dx.doi.org/10.2174/1874609808666150702095803] [PMID: 26135528]

[28]   Szalay MS, Kovács IA, Korcsmáros T, Böde C, Csermely P. Stress-induced rearrangements of cellular networks: Consequences for protection and drug design. FEBS Lett 2007; 581(19): 3675-80.
[http://dx.doi.org/10.1016/j.febslet.2007.03.083] [PMID: 17433306]

[29]   Jeong H, Tombor B, Albert R, Oltvai ZN, Barabási AL. The large-scale organization of metabolic networks. Nature 2000; 407(6804): 651-4.
[http://dx.doi.org/10.1038/35036627] [PMID: 11034217]

[30]   Calabrese V, Cornelius C, Dinkova-Kostova AT, Calabrese EJ, Mattson MP. Cellular stress responses, the hormesis paradigm, and vitagenes: novel targets for therapeutic intervention in neurodegenerative disorders. Antioxid Redox Signal 2010; 13(11): 1763-811.
[http://dx.doi.org/10.1089/ars.2009.3074] [PMID: 20446769]

[31]   Rattan SI, Gonzalez-Dosal R, Nielsen ER, Kraft DC, Weibel J, Kahns S. Slowing down aging from within: mechanistic aspects of anti-aging hormetic effects of mild heat stress on human cells. Acta Biochim Pol 2004; 51(2): 481-92.
[PMID: 15218544]

[32]   Barnett SA. Exploratory behaviour. Br J Psychol 1958; 49(4): 289-310.
[http://dx.doi.org/10.1111/j.2044-8295.1958.tb00667.x] [PMID: 13596571]

[33]   Brusch H, Ghoshal S. A Bias for action. Boston, MA: Harvard Business School Press 2004.

[34]   (Mather E. Novelty, attention, and challenges for developmental psychology. Front Psychol 2013; 01 [http://dx.doi.org/10.3389/fpsyg.2013.00491]

[35]   Rattan SI. Rationale and methods of discovering hormetins as drugs for healthy ageing. Expert Opin Drug Discov 2012; 7(5): 439-48.
       [http://dx.doi.org/10.1517/17460441.2012.677430] [PMID: 22509769]

[36]   da Costa AN, Feitosa JV, Montezuma PA Jr, de Souza PT, de Araújo AA. Rectal temperatures, respiratory rates, production, and reproduction performances of crossbred Girolando cows under heat stress in northeastern Brazil. Int J Biometeorol 2015; 59(11): 1647-53.
       [http://dx.doi.org/10.1007/s00484-015-0971-4] [PMID: 25702060]

[37]   Ruebush   M.   Dirt,   Germs,   and   Other   Friendly   Filth.   Available   from: https://experiencelife.com/article/dirt-germs-and-other-friendly-filth/ 2012.

[38]   Calabrese EJ. Historical foundations of wound healing and its potential for acceleration: dose-response considerations. Wound Repair Regen 2013; 21(2): 180-93.
       [http://dx.doi.org/10.1111/j.1524-475X.2012.00842.x] [PMID: 23421727]

[39]   Vadyvaloo V, Jarrett C, Sturdevant DE, Sebbane F, Hinnebusch BJ. Transit through the flea vector induces a pretransmission innate immunity resistance phenotype in Yersinia pestis. PLoS Pathog 2010; 6(2): e1000783.
       [http://dx.doi.org/10.1371/journal.ppat.1000783] [PMID: 20195507]

[40]   Cox J, Mota J, Sukupolvi-Petty S, Diamond MS, Rico-Hesse R. Mosquito bite delivery of dengue virus enhances immunogenicity and pathogenesis in humanized mice. J Virol 2012; 86(14): 7637-49.
       [http://dx.doi.org/10.1128/JVI.00534-12] [PMID: 22573866]

[41]   Palm NW, Rosenstein RK, Yu S, Schenten DD, Florsheim E, Medzhitov R. Bee venom phospholipase A2 induces a primary type 2 response that is dependent on the receptor ST2 and confers protective immunity. Immunity 2013; 39(5): 976-85.
       [http://dx.doi.org/10.1016/j.immuni.2013.10.006] [PMID: 24210353]

[42]   Sarr JB, Samb B, Sagna AB, *et al.* Differential acquisition of human antibody responses to Plasmodium falciparum according to intensity of exposure to Anopheles bites. Trans R Soc Trop Med Hyg 2012; 106(8): 460-7.
       [http://dx.doi.org/10.1016/j.trstmh.2012.05.006] [PMID: 22721883]

[43]   Jones L. Thermal Touch. Scholarpedia 2009; 4(5): 7955.
       [http://dx.doi.org/10.4249/scholarpedia.7955]

[44]   van den Oord QG, van Wijk EJ, Lugton IW, Morris RS, Holmes CW. Effects of air temperature, air movement and artificial rain on the heat production of brushtail possums (Trichosurus vulpecula): an exploratory study. N Z Vet J 1995; 43(7): 328-32.
       [http://dx.doi.org/10.1080/00480169./1995.35914] [PMID: 16031874]

[45]   Carroll JA, Burdick NC, Chase CC Jr, Coleman SW, Spiers DE. Influence of environmental temperature on the physiological, endocrine, and immune responses in livestock exposed to a provocative immune challenge. Domest Anim Endocrinol 2012; 43(2): 146-53.
       [http://dx.doi.org/10.1016/j.domaniend.2011.12.008] [PMID: 22425434]

[46]   Rattan SI, Fernandes RA, Demirovic D, Dymek B, Lima CF. Heat stress and hormetin-induced hormesis in human cells: effects on aging, wound healing, angiogenesis, and differentiation. Dose Response 2009; 7(1): 90-103.
[http://dx.doi.org/10.2203/dose-response.08-014.Rattan] [PMID: 19343114]

[47]   Weisser B. [Sauna bathing is beneficial for cardiac health-the more, the better [corrected]]. Dtsch Med Wochenschr 2015; 140(13): 970-1.
[PMID: 26115128]

[48]   Krause M, Ludwig MS, Heck TG, Takahashi HK. Heat shock proteins and heat therapy for type 2 diabetes: pros and cons. Curr Opin Clin Nutr Metab Care 2015; 18(4): 374-80.
[http://dx.doi.org/10.1097/MCO.0000000000000183] [PMID: 26049635]

[49]   Tomiyama C, Watanabe M, Honma T, *et al.* The effect of repetitive mild hyperthermia on body temperature, the autonomic nervous system, and innate and adaptive immunity. Biomed Res 2015; 36(2): 135-42.
[http://dx.doi.org/10.2220/biomedres.36.135] [PMID: 25876664]

[50]   Laukkanen T, Khan H, Zaccardi F, Laukkanen JA. Association between sauna bathing and fatal cardiovascular and all-cause mortality events. JAMA Intern Med 2015; 175(4): 542-8.
[http://dx.doi.org/10.1001/jamainternmed.2014.8187] [PMID: 25705824]

[51]   Gryka D, Pilch W, Szarek M, Szygula Z, Tota Ł. The effect of sauna bathing on lipid profile in young, physically active, male subjects. Int J Occup Med Environ Health 2014; 27(4): 608-18.
[http://dx.doi.org/10.2478/s13382-014-0281-9] [PMID: 25001587]

[52]   Crinnion WJ. Sauna as a valuable clinical tool for cardiovascular, autoimmune, toxicant- induced and other chronic health problems. Altern Med Rev 2011; 16(3): 215-25.
[PMID: 21951023]

[53]   Heylighen F. Evolutionary Well-Being: the paleolithic model. Available from: http://ecco.vub.ac.be/?q=node/127 2010.

[54]   Manheimer EW, van Zuuren EJ, Fedorowicz Z, Pijl H. Paleolithic nutrition for metabolic syndrome: systematic review and meta-analysis. Am J Clin Nutr 2015; 102(4): 922-32.
[http://dx.doi.org/10.3945/ajcn.115.113613] [PMID: 26269362]

[55]   Eaton SB, Eaton SB. An evolutionary perspective on human physical activity: implications for health. Comp Biochem Physiol A Mol Integr Physiol 2003; 136(1): 153-9.
[http://dx.doi.org/10.1016/S1095-6433(03)00208-3] [PMID: 14527637]

[56]   Jew S, AbuMweis SS, Jones PJ. Evolution of the human diet: linking our ancestral diet to modern functional foods as a means of chronic disease prevention. J Med Food 2009; 12(5): 925-34.
[http://dx.doi.org/10.1089/jmf.2008.0268] [PMID: 19857053]

[57]   Taormina G, Mirisola MG. Longevity: epigenetic and biomolecular aspects. Biomol Concepts 2015; 6(2): 105-17.
[http://dx.doi.org/10.1515/bmc-2014-0038] [PMID: 25883209]

[58]   Honjoh S, Yamamoto T, Uno M, Nishida E. Signalling through RHEB-1 mediates intermittent fasting-induced longevity in C. elegans. Nature 2009; 457(7230): 726-30.
[http://dx.doi.org/10.1038/nature07583] [PMID: 19079239]

[59]   Cherif A, Roelands B, Meeusen R, Chamari K. Effects of Intermittent Fasting, Caloric Restriction, and Ramadan Intermittent Fasting on Cognitive Performance at Rest and During Exercise in Adults. Sports Med 2015. [Epub ahead of print].
[PMID: 26438184]

[60]   Mattson MP. Challenging oneself intermittently to improve health. Dose Response 2014; 12(4): 600-18.
[http://dx.doi.org/10.2203/dose-response.14-028.Mattson] [PMID: 25552960]

[61]   Wilhelm M, Zueger T, De Marchi S, *et al.* Inflammation and atrial remodeling after a mountain marathon. Scand J Med Sci Sports 2014; 24(3): 519-25.
[http://dx.doi.org/10.1111/sms.12030] [PMID: 23253265]

[62]   O'Keefe JH, Vogel R, Lavie CJ, Cordain L. Exercise like a hunter-gatherer: a prescription for organic physical fitness. Prog Cardiovasc Dis 2011; 53(6): 471-9.
[http://dx.doi.org/10.1016/j.pcad.2011.03.009] [PMID: 21545934]

[63]   Schwabe K, Schwellnus M, Derman W, Swanevelder S, Jordaan E. Medical complications and deaths in 21 and 56 km road race runners: a 4-year prospective study in 65 865 runners--SAFER study I. Br J Sports Med 2014; 48(11): 912-8.
[http://dx.doi.org/10.1136/bjsports-2014-093470] [PMID: 24735839]

[64]   Hsieh SY, Hsu CY, He JR, *et al.* Identifying apoptosis-evasion proteins/pathways in human hepatoma cells *via* induction of cellular hormesis by UV irradiation. J Proteome Res 2009; 8(8): 3977-86.
[http://dx.doi.org/10.1021/pr900289g] [PMID: 19545154]

[65]   Sokolov M, Nguyen V, Neumann R. Comparative analysis of whole-genome gene expression changes in cultured human embryonic stem cells in response to low, clinical diagnostic relevant, and high doses of ionizing radiation exposure. Int J Mol Sci 2015; 16(7): 14737-48.
[http://dx.doi.org/10.3390/ijms160714737] [PMID: 26133243]

[66]   Le Bourg É. [Mild stress as a means to modulate aging: from fly to human?]. Med Sci (Paris) 2012; 28(3): 305-10.
[http://dx.doi.org/10.1051/medsci/2012283019] [PMID: 22480655]

[67]   Sesti-Costa R, Ignacchiti MD, Chedraoui-Silva S, Marchi LF, Mantovani B. Chronic cold stress in mice induces a regulatory phenotype in macrophages: correlation with increased 11β-hydroxysteroid dehydrogenase expression. Brain Behav Immun 2012; 26(1): 50-60.
[http://dx.doi.org/10.1016/j.bbi.2011.07.234] [PMID: 21801831]

[68]   Dirt,   Germs,   and   Other   Friendly   Filth.   Available   from:
http://science.nasa.gov/science-news/science-at-nasa/2003/07feb_stronggravity/ 2003.

[69]   Minois N. The hormetic effects of hypergravity on longevity and aging. Dose Response 2006; 4(2): 145-54.
[http://dx.doi.org/10.2203/dose-response.05-008.Minois] [PMID: 18648640]

[70]   Rattan SI. Aging intervention, prevention, and therapy through hormesis. J Gerontol A Biol Sci Med Sci 2004; 59(7): 705-9.
[http://dx.doi.org/10.1093/gerona/59.7.B705] [PMID: 15304535]

[71]   Tajino J, Ito A, Nagai M, *et al.* Intermittent application of hypergravity by centrifugation attenuates

disruption of rat gait induced by 2 weeks of simulated microgravity. Behav Brain Res 2015; 287(287): 276-84.
[http://dx.doi.org/10.1016/j.bbr.2015.03.030] [PMID: 25819803]

[72]   Iwasaki K, Ogawa Y, Aoki K, Yanagida R. Cerebral circulation during mild +Gz hypergravity by short-arm human centrifuge. J Appl Physiol 2012; 112(2): 266-71.
[http://dx.doi.org/10.1152/japplphysiol.01232.2011] [PMID: 22052869]

[73]   Cutler GC, Rix RR. Can poisons stimulate bees? Appreciating the potential of hormesis in bee-pesticide research. Pest Manag Sci 2015; 71(10): 1368-70.
[http://dx.doi.org/10.1002/ps.4042] [PMID: 25989135]

[74]   Diaz GJ, Calabrese E, Blain R. Aflatoxicosis in chickens (*Gallus gallus*): an example of hormesis? Poult Sci 2008; 87(4): 727-32.
[http://dx.doi.org/10.3382/ps.2007-00403] [PMID: 18339995]

[75]   Yang H, Mao Y, Tan B, Luo S, Zhu Y. The protective effects of endogenous hydrogen sulfide modulator, S-propargyl-cysteine, on high glucose-induced apoptosis in cardiomyocytes: A novel mechanism mediated by the activation of Nrf2. Eur J Pharmacol 2015; 761: 135-43.
[http://dx.doi.org/10.1016/j.ejphar.2015.05.001] [PMID: 25979858]

[76]   Zhang YX, Hu KD, Lv K, *et al.* 2015.The hydrogen sulfide donor nahs delays programmed cell death in barley aleurone layers by acting as an antioxidant. Oxid Med Cell Longev 2015.
[http://dx.doi.org/10.1155/2015/714756]

[77]   Yang G, An SS, Ji Y, Zhang W, Pei Y. Hydrogen Sulfide Signaling in Oxidative Stress and Aging Development. Oxid Med Cell Longev 2015; 2015: 357824.
[http://dx.doi.org/10.1155/2015/357824] [PMID: 26075033]

[78]   Tran BH, Huang C, Zhang Q, *et al.* Cardioprotective effects and pharmacokinetic properties of a controlled release formulation of a novel hydrogen sulfide donor in rats with acute myocardial infarction. Biosci Rep 2015; 35(3): e00216.
[PMID: 26182378]

[79]   Kyriazis M. Nonlinear stimulation and hormesis in human aging: practical examples and action mechanisms. Rejuvenation Res 2010; 13(4): 445-52.
[http://dx.doi.org/10.1089/rej.2009.0996] [PMID: 20662589]

[80]   Hirsch HV, Mercer J, Sambaziotis H, *et al.* Behavioral effects of chronic exposure to low levels of lead in Drosophila melanogaster. Neurotoxicology 2003; 24(3): 435-42.
[http://dx.doi.org/10.1016/S0161-813X(03)00021-4] [PMID: 12782108]

[81]   López-Martínez G, Hahn DA. Early life hormetic treatments decrease irradiation-induced oxidative damage, increase longevity, and enhance sexual performance during old age in the Caribbean fruit fly. PLoS One 2014; 9(1): e88128.
[http://dx.doi.org/10.1371/journal.pone.0088128] [PMID: 24498251]

[82]   Liu SZ. Radiation hormesis. A new concept in radiological science. Chin Med J (Engl) 1989; 102(10): 750-5.
[PMID: 2517054]

[83]   Gajecka M. The effect of low-dose experimental zearalenone intoxication on the immunoexpression of

estrogen receptors in the ovaries of pre-pubertal bitches. Pol J Vet Sci 2012; 15(4): 685-91.
[PMID: 23390758]

[84]   Zhang H, Liu B, Zhou Q, *et al.* Alleviation of pre-exposure of mouse brain with low-dose 12C6+ ion
         or 60Co gamma-ray on male reproductive endocrine damages induced by subsequent high-dose
         irradiation. Int J Androl 2006; 29(6): 592-6.
         [http://dx.doi.org/10.1111/j.1365-2605.2006.00698.x] [PMID: 17121657]

[85]   Sohrabji F, Lewis DK. Estrogen-BDNF interactions: implications for neurodegenerative diseases.
         Front Neuroendocrinol 2006; 27(4): 404-14.
         [http://dx.doi.org/10.1016/j.yfrne.2006.09.003] [PMID: 17069877]

[86]   Peters A. Testosterone and carotenoids: an integrated view of trade-offs between immunity and sexual
         signalling. BioEssays 2007; 29(5): 427-30.
         [http://dx.doi.org/10.1002/bies.20563] [PMID: 17450573]

[87]   Incrocci L, Hop WC, Slob AK. Visual erotic and vibrotactile stimulation and intracavernous injection
         in screening men with erectile dysfunction: a 3 year experience with 406 cases. Int J Impot Res 1996;
         8(4): 227-32.
         [PMID: 8981172]

[88]   Rowland DL, Incrocci L, Slob AK. Aging and sexual response in the laboratory in patients with
         erectile dysfunction. J Sex Marital Ther 2005; 31(5): 399-407.
         [http://dx.doi.org/10.1080/00926230591006520] [PMID: 16169823]

[89]   Rowland DL. Genital and heart rate response to erotic stimulation in men with and without premature
         ejaculation. Int J Impot Res 2010; 22(5): 318-24.
         [http://dx.doi.org/10.1038/ijir.2010.22] [PMID: 20861844]

[90]   Huynh HK, Beers C, Willemsen A, *et al.* High-intensity erotic visual stimuli de-activate the primary
         visual cortex in women. J Sex Med 2012; 9(6): 1579-87.
         [http://dx.doi.org/10.1111/j.1743-6109.2012.02706.x] [PMID: 22489578]

[91]   Faccio E, Casini C, Cipolletta S. Forbidden games: the construction of sexuality and sexual pleasure
         by BDSM 'players'. Cult Health Sex 2014; 16(7): 752-64.
         [http://dx.doi.org/10.1080/13691058.2014.909531] [PMID: 24828811]

[92]   Reynolds GL, Fisher DG, Rogala B. Why women engage in anal intercourse: results from a qualitative
         study. Arch Sex Behav 2015; 44(4): 983-95.
         [http://dx.doi.org/10.1007/s10508-014-0367-2] [PMID: 25378264]

[93]   Rehor JE. Sensual, erotic, and sexual behaviors of women from the "kink" community. Arch Sex
         Behav 2015; 44(4): 825-36.
         [http://dx.doi.org/10.1007/s10508-015-0524-2] [PMID: 25795531]

[94]   Heylighen F. Challenge Propagation: Towards a theory of distributed intelligence and the global brain.
         Spanda Journal 2014; V: 2.

[95]   Shannon CE, Weaver W. The mathematical theory of communication. Urbana, IL: University of
         Illinois press 1949.

[96]   Shannon CE. A mathematical theory of communication. Bell System Technical Journal 1948; 27: 379-
         115.

[97]   Beigi S. A Unifying Theory of Resilience for Volatile, Uncertain, Complex and Ambiguous (VUCA) World. PhD Thesis, University of Bristol: UK, 2014.

[98]   Fonseca R, Blascovich J, Garcia-Marques T. Challenge and threat motivation: effects on superficial and elaborative information processing. Front Psychol 2014; 5: 1170.
[http://dx.doi.org/10.3389/fpsyg.2014.01170] [PMID: 25352823]

[99]   Blascovich J. Challenge and threat. In: Elliot AJ, Ed. Handbook of Approach and Avoidance Motivation. New York: Psychology Press 2008; pp. 431-45.

[100]   Fukudo S, Kanazawa M. Gene, environment, and brain-gut interactions in irritable bowel syndrome. J Gastroenterol Hepatol 2011; 26 (Suppl. 3): 110-5.
[http://dx.doi.org/10.1111/j.1440-1746.2011.06631.x] [PMID: 21443722]

[101]   Tops M, Boksem MA, Koole SL. Subjective effort derives from a neurological monitor of performance costs and physiological resources. Behav Brain Sci 2013; 36(6): 703-4.
[http://dx.doi.org/10.1017/S0140525X13001167] [PMID: 24304802]

[102]   Prévost C, Pessiglione M, Météreau E, Cléry-Melin ML, Dreher JC. Separate valuation subsystems for delay and effort decision costs. J Neurosci 2010; 30(42): 14080-90.
[http://dx.doi.org/10.1523/JNEUROSCI.2752-10.2010] [PMID: 20962229]

[103]   Hennecke M, Freund AM. Competing goals draw attention to effort, which then enters cost-benefit computations as input. Behav Brain Sci 2013; 36(6): 690-1.
[http://dx.doi.org/10.1017/S0140525X13001027] [PMID: 24304788]

[104]   Tay KH, Luan Q, Croft A, *et al.* Sustained IRE1 and ATF6 signaling is important for survival of melanoma cells undergoing ER stress. Cell Signal 2014; 26(2): 287-94.
[http://dx.doi.org/10.1016/j.cellsig.2013.11.008] [PMID: 24240056]

[105]   Naughton MC, McMahon JM, FitzGerald U. Differential activation of ER stress pathways in myelinating cerebellar tracts. Int J Dev Neurosci 2015.
[http://dx.doi.org/10.1016/j.ijdevneu.2015.08.002]

[106]   Locke M, Celotti C. The effect of heat stress on skeletal muscle contractile properties. Cell Stress Chaperones 2014; 19(4): 519-27.
[http://dx.doi.org/10.1007/s12192-013-0478-z] [PMID: 24264930]

[107]   Taleb NN. Antifragile: Things That Gain from Disorder. New York: Random House 2012.

[108]   Heylighen F. The science of self-organisation and adaptivity The encyclopaedia of life support systems. Oxford: Eolss Publishers 2001.

[109]   Yerkes RM, Dodson JD. The relation of strength of stimulus to rapidity of habit-formation. J Comp Neurol Psychol 1908; 18: 459-82.
[http://dx.doi.org/10.1002/cne.920180503]

[110]   The Temporal Dynamics Model of Emotional Memory Processing: A Synthesis on the Neurobiological Basis of Stress-Induced Amnesia, Flashbulb and Traumatic Memories, and the Yerkes-Dodson Law. Neural Plasticity 2007; 33

[111]   Csikszentmihalyi M, Rathunde K. The measurement of flow in everyday life: toward a theory of emergent motivation. Nebr Symp Motiv 1992; 40: 57-97.

[PMID: 1340523]

[112] Csikszentmihalyi M. Flow: The Psychology of Optimal Experience. New York, NY: Harper and Row 1990.

[113] Blascovich J, Mendes WB. Challenge and threat appraisals: The role of affective cues. In: Forgas J, Ed. Feeling and thinking: The role of affect in social cognition. Cambridge, UK: Cambridge University Press 2000; pp. 59-82.

[114] Kurzban R, Duckworth A, Kable JW, Myers J. An opportunity cost model of subjective effort and task performance. Behav Brain Sci 2013; 36(6): 661-79.
[http://dx.doi.org/10.1017/S0140525X12003196] [PMID: 24304775]

[115] Hofmann W, Kotabe H. On treating effort as a dynamically varying cost input. Behav Brain Sci 2013; 36(6): 692-3.
[http://dx.doi.org/10.1017/S0140525X13001040] [PMID: 24304790]

[116] Tang YY, Lu Q, Feng H, Tang R, Posner MI. Short-term meditation increases blood flow in anterior cingulate cortex and insula. Front Psychol 2013; 6(212)
[http://dx.doi.org/10.3389/fpsyg.2015.00212]

[117] Leung MK, Chan CC, Yin J, Lee CF, So KF, Lee TM. Enhanced amygdala-cortical functional connectivity in meditators. Neurosci Lett 2015; 590: 106-10.
[http://dx.doi.org/10.1016/j.neulet.2015.01.052] [PMID: 25623035]

[118] Badowska-Szalewska E, Ludkiewicz B, Spodnik JH, Krawczyk R, Moryś J. The influence of mild stressors on neurons containing interleukin-1β in the central (CeA) and medial (MeA) amygdala in the ageing process of rats. Acta Neurobiol Exp (Warsz) 2015; 75(3): 279-92.
[PMID: 26581384]

[119] Saez A, Rigotti M, Ostojic S, Fusi S, Salzman CD. Abstract Context Representations in Primate Amygdala and Prefrontal Cortex. Neuron 2015; 87(4): 869-81.
[http://dx.doi.org/10.1016/j.neuron.2015.07.024] [PMID: 26291167]

[120] Milisav I, Poljsak B, Suput D. Adaptive response, evidence of cross-resistance and its potential clinical use. Int J Mol Sci 2012; 13(9): 10771-806.
[http://dx.doi.org/10.3390/ijms130910771] [PMID: 23109822]

CHAPTER 2

Vitagenes and Hormetins: The Pills of Hormesis

Abstract: In the previous chapter I discussed the physiological basis of hormesis and given examples of how certain practical actions can have a hormetic benefit upon our health. Here I will examine examples of oral drugs, compounds or supplements which may exhibit a hormetic effect. These compounds are called hormetins and many of these act on vitagenes, genes which encode transcription factors that are necessary for maintaining health. Thus, I explore a direct link between **information** carried by chemical compounds, and **physiological modulation**, with resulting health improvement. Some of these hormetins are pluripotent agents, exhibiting benefits at multiple levels and tissues. Other compounds are mimics of hormetic physical actions such as calorie restriction and exercise. I will also discuss the phenomenon of xenohormesis, *i.e.* hormetic gains experienced by humans through hormetins which originate from a different species. The discussion here complements both the concept of hormesis and that of Environmental Enrichment which will be discussed in the next chapter, and the aim is to provide a comprehensive approach ensuring a better understanding of hormetic mechanisms in a wider sense.

Keywords: Adaptation, Calorie restriction mimetics, Exercise mimetics, Vitagene, Hormesis, Hormetic pathways, Hormetin, Inter-individual variations, Sex mimetics, Xenohormesis.

INTRODUCTION

Hormesis affects many age-related processes [1 - 3]. However, the physiological mechanisms of the hormetic action itself may also be mimicked by certain chemical compounds (such as the active ingredients [4] extracted from the roots of the Chinese herb Sanchi (*Panax notoginseng*) for example), which act to up-regulate signalling and other pathways sharing the same physiological mechanisms as the hormetic intervention itself [5, 6]. These agents are called hormetins [7]. Hormetins may act upon vitagenes [8], a group of genes which are

Marios Kyriazis

involved, through a complex mechanism, in maintaining the health of cells during ageing. Vitagenes encode proteins (such as Heat Shock Proteins –Hsp) which are normally expressed after a stressful episode, and also encode the sirtuin protein system which is involved in ageing [9, 10] (Fig. **1**).

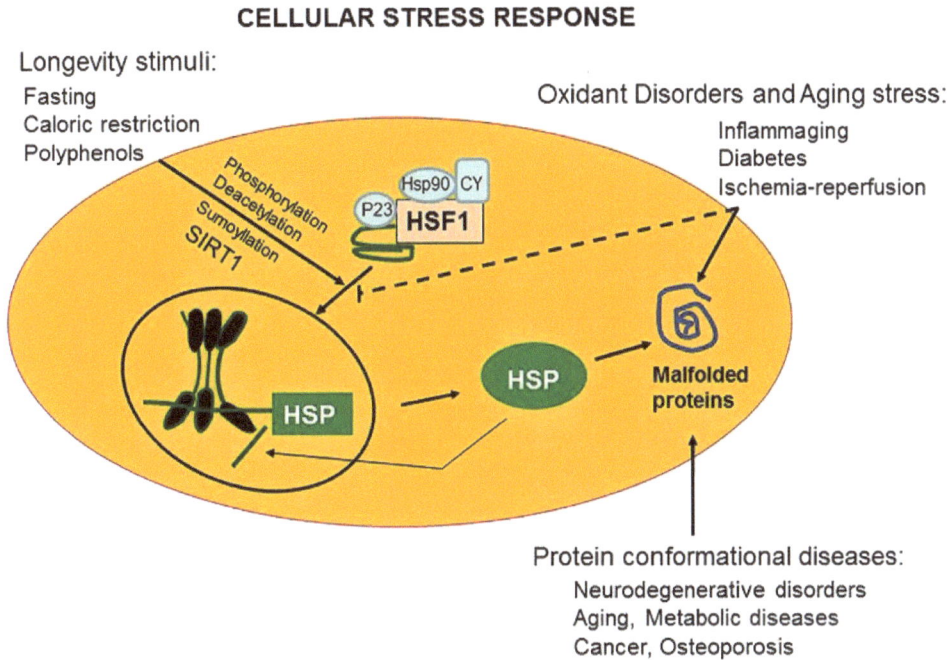

Fig. (1). Vitagenes and the pathway of cellular stress response. Misfolded proteins cumulating in response to proteotoxic stresses trigger the cellular stress response. Hsps that are normally bound to HSF1, maintaining it in a repressed state before stress, are titrated away by damaged or misfolded proteins with resulting HSF-1 activation. Multi-step activation of HSF1 involves post-translational modifications, such as hyperphosphorylation, deacetylation, or sumoylation... Nutritional anti-oxidants, are able to activate vitagenes, such as heme oxygenase, Hsp70, thioredoxin reductase and sirtuins which represent an integrated system for cellular stress tolerance. Activation of vitagene system, with up-regulation of HO-1, thioredoxin, GSH, and sirtuin, results in reduction of pro-oxidant conditions. During inflammaging, including aged-associated pathologies, such as Alzheimer's disease and osteoporosis, a gradual decline in potency of the heat shock response occur and this may prevent repair of protein damage, leading to degeneration and cell death of critical parenchymal cells. Image and text credit from [11].

The concept of hormetins shares common frontiers with that of calorie restriction, in the sense that many mechanisms involved are similar or shared. This suggests that it is possible to choose a range of compounds or treatments which have pluripotent effects for a maximum overall benefit.

It is known that high amounts of polyphenols found in the Mediterranean diet act along hormetic pathways and regulate stress resistance. Pathways involved include those that inhibit mTOR (mechanical Target of Rapamycin) and those which modulate FOXO (Forkhead box O genes) [12]. This indicates that compounds which influence stress resistance (hormetins) may have actions similar to mTOR inhibitors or to other mimetics (discussed below).

Inter-Individual Variations

Before I discuss some of these hormetins I must highlight that there is considerable inter (and even intra-)-individual variation in the response to stress. Individuals who have similar phenotypes at rest may react differently when exposed to a stressful challenge. This concept is very important in the entire discussion about hormesis and hormetic lifestyle [13]. The implication of this is that there is still a lot to learn about the hormetic mechanisms, which could be expressed differently, even in the same individuals under different conditions. The long-time hormesis authority Edward Calabrese [14] quotes:

*A principal concern in assessing the effects of drugs on humans is that of inter-individual variation. Numerous factors are known that contribute to such variation, including age, familial background, gender, nutritional status, the presence of pre-existing disease, amongst other factors. Using the hormesis database we ... identified a substantial number of experimental settings in which hormesis had been studied in individuals or closely related strains of organisms which differed in susceptibility to toxic agents. In these evaluations we compared responses where the range in susceptibility varied **from less than 10 fold to well in excess of 100 fold.** Of particular note was that the hormetic response was generally independent of susceptibility, with hormetic responses occurring in subjects ranging from high to low susceptibility. Likewise, the quantitative features of the hormetic dose–response are independent of susceptibility. In about 20% of the cases, it appeared that the lack of an hormetic response in a susceptible strain or subgroup was related to its* *increased risk. These observations have important implications in the development of treatment strategies for patients* (emphasis mine).

When calculating the optimal dose for any hormetic compound, even for an average individual, there could be a greater than 20-fold variation in the safety profile, so there could be a range of possible responses to any given hormetin [15]. The dose may overshoot the optimal range where the benefit could be found, and thus result in toxicity. This overlapping of the optimal dose landscape with that of toxicity will necessitate a continual monitoring of the effects of hormetins on any specific individual. Therefore, it is important to account for the variations originating from any aspect of the organism's environment. This is, of course, a difficult or impossible feat, but nevertheless the notion adds to the validity of the general hormetic principle, which is based on variation, novelty, change, unpredictability, non-linearity, and subsequent adaptation. In this case, 'unpredictability' refers to a situation whereby the dose needs to be determined empirically due to environmental and genetic variability to susceptibility.

Hormetins: Examples and Actions

It is necessary to highlight once more that hormetins act through a dose-response phenomenon. The dose of the challenging stimulus is important. Too high a dose may result in the opposite effect compared to a low/mild dose.

Assuming that we have accounted for inter-individual variations and we have succeeded in remaining within the homeodynamic space (where the organism functions optimally having been subjected to an optimal dose of the stimulus), I will now mention some examples of hormetins, discrete chemical agents which modulate hormesis.

a. Metformin

This biguanide is used primarily in diabetes but it also has hormetic effects (as well as being a calorie restriction mimetic, as discussed below). It is known that metformin can prolong lifespan in experimental animals [16]. More recently, the hormetic basis of metformin has been further elucidated. It is thought that metformin causes a mild injury following elevation of reactive oxygen species (ROS) in the mitochondria which activates repair mechanisms and subsequently improves overall function [17]. In addition, metformin acts by modulating the stress response by suppressing the AMPK/mTOR/S6K1 axis, as well as suppressing protein kinases and protein kinase receptors [18].

b. Curcumin

Apart from its several other health benefits, curcumin (diferuloylmethane) is also classified as a hormetin [19]. One of its actions is to modulate stress response pathways. Curcumin stimulates synthesis of heat shock proteins and regulates proteasome function. As a result, the response to mild stress is controlled and this activates cell repair mechanisms [20]. It can improve wound healing acting in a hormetic manner (low dose improves wound healing whereas higher doses inhibits it) [21]. In a characteristic study, Demirovic and Rattan [22] conclude:

> *"...curcumin, a component of the widely used spice turmeric, modulates wound healing in vitro in a biphasic dose response manner, being stimulatory at low doses (between 1 and 5 μM), and inhibitory at higher doses. Furthermore, our results show that the hormetic effects of low levels of curcumin are achieved by virtue of it being a hormetin in terms of the induction of stress response pathways, including Nrf2 and HO-1 in human cells".*

c. Carnosine

Carnosine is a pluripotent dipeptide which has a wide range of actions as an antioxidant and anti-glycator [23]. Carnosine can also act as a calorie restriction mimetic as described below. In addition, it is known that carnosine activates vitagenes involved in hormesis which enhance cellular defences against external and internal poisons [24]. Carnosine shares several characteristics of hormetins: it upregulates DNA repair following stress, improves macromolecular repair following free radical damage, up-regulates immune components and regulates protein cross-linking [25].

Other hormetic-based activities have been described, both for carnosine itself and for its metabolites. For example, carnosine and its component, β-alanine control stress response and hypothalamic hormones which are regulators of the metabolism [26]. Furthermore, during physical exercise (which is a form of hormetic, mildly stressful event) there is release of carnosine which further regulates physiological functions [27].

d. Acetyl-L-carnitine

This is another mild hormetin which may modulate the stress response. Acetyl--carnitine (ALC) triggers the expression of vitagenes, which then modulate heat shock proteins, that initiate several other processes involved in the stress response and protect the cell during stress. These processes repair any damage, including age-related damage, and not just the damage caused by the stressful event itself. As a final result, the functionality of the cell is maintained. The mechanisms involved in this particular example of ALC have been studied in some detail. It was reported that ALC modulates several signalling pathways which result in elimination of cancer cells, where at the same time, enhance repair of normal cells [28]. I must highlight that the function of heat shock proteins declines with age and if this is left untreated, it may lead to inappropriate responses to stress, where even mild stress can cause cellular damage. Therefore, in this respect, ALC is a compound which, through modulation of heat shock protein function, can lower the risk of this from happening [9].

e. Artemisinin

Research shows that artemisinin is a calorie restriction mimetic [29] but it can also be classified as a hormetin, a typical example of how the two processes overlap. Artemisinin is used in traditional Chinese medicine and it is an extract of the plant sweet wormwood (*Artemisia annua*). It is an effective anti-malarial compound, and it is now produced using genetically engineered yeast [30]. Both artemisinin and its derivatives (for example, the semi-synthetic artesunate) have been studied in situations involving nitric oxide. Artemisinin mimics the beneficial effects of nitric oxide (and those of hydrogen peroxide, another agent found to improve health parameters) and it is therefore considered as a lifespan extension agent, at least in yeast [31].

f. Kinetin

N6-furfuryladenine (kinetin) is a hormetin which has been shown to delay age-related damage in human fibroblasts. Kinetin is a natural plant growth hormone, which can be also found in human cells [32]. It also prolongs lifespans of insects and protects against senescence in plants, but importantly, it was also shown to reduce fecundity [33]. This aligns well with the notion I will discuss later in the book, whereby there are clear trade-offs between improved somatic maintenance and reproduction (germ line maintenance). Kinetin modulates cellular ageing in a hormetic manner (by activating stress response pathways and stress proteins such as Hsp90 and Hsp70). This effect is seen during heat stress and also with exposure to another hormetin, curcumin as discussed above [34].

Other Hormetins

* Ferulic Acid

Found in maize and vegetables, ferulic acid is a polyphenol which scavenges free radicals. Barone *et al.* [35] have proposed it as a possible hormetin, as it was shown to induce stress response by up-regulating caspases, NO-synthetase, heme oxygenase-1, Hsp70, and extracellular signal-regulated kinase. Many of these stress response mechanisms are also activated by other hormetins. In other words, several hormetins share common stress response modulating actions.

* DADLE

D-Ala(2)-D-Leu(5)-Enkephalin (also called dalargin) is a peptide which regulates the delta opioid peptide receptor. This particular receptor is active during hibernation of some animals and is involved with the regulation of muscle damage following ischaemia, oxygen and oxidation metabolism, and protection against neurotoxicity [36]. DADLE is therefore a 'cold-mimetic hormetin', *i.e.* if replicates the effects of cold stress on the body [37].

* Silymarin

The extracts of the seeds of the milk thistle (*Silybum marianum*) have been used in a variety of situations involving liver disease, as an antioxidant and as a general 'anti-ageing' supplement. It has hormetic (dose response) effects which have been studied in some detail in C. elegans [38]. Silymarin modulates stress response in arsenic toxicity and reduces the expression of Hsp70, lipid peroxidation and other inflammation markers [39].

* Geranylgeranylacetone (GGA)

GGA is a naturally-occurring compound found in herbs, and has been used for many years as an anti-ulcer drug. It induces apoptosis of certain cancer cells, but is also a heat-mimetic: it induces production of heat stress proteins [40]. However, in true hormetic spirit, while low doses are clearly beneficial, high doses are toxic [41].

* Oltipraz

This organosulfur compound is effective against schistosomiasis, but it has also been shown to inhibit the formation of several types of cancers in rodents [42]. Unfortunately, perhaps due to a hormetic effect, higher doses are toxic and research in its benefits has lost momentum. However, it is known that low doses can modulate several biochemical pathways which are involved in protecting against oxidation damage [43].

* Naringenin

This is a bioflavonoid found in relatively large amounts in grapefruit and tomatoes. As in the case of many bioflavonoids, naringenin may reduce inflammatory markers, oestrogen levels, blood lipids and cancer. Flies treated with lower doses of narigenin showed resistance to stress and an increased lifespan, whereas those treated with higher doses, had a shorter lifespan, or where immediately poisoned [44]. Here again we encounter the phenomenon whereby increased longevity is associated with decrease fecundity.

Calorie Restriction, and Calorie Restriction Mimetics in Nutritional Hormesis

As mentioned previously, a hormetic effect can be achieved through nutritional challenges. Here, I am referring to techniques such as Calorie Restriction (CR), and intermittent Fasting (IF). These techniques challenge the organism to respond to the stressful nutritional event, and modify a host of physiological and pathological parameters. It is known that CR, as well as exercise (see below), can affect apoptosis and cellular remodelling, among other effects [45]. In order to gain a more comprehensive picture of the effects of calorie restriction, it will be necessary to consider its physiological consequences in three broadly distinct groups:

* genetic (regulation of genes such as Sir-2, Daf-16 and p53 all of which have direct relevance to age-related dysfunction)

* biochemical (modulation of DHEA-dehydroepiandrosterone, inflammation response, IGF-1, nerve growth factors, and mitochondrial function among others), and

* physical (reduction of cholesterol, lipids, improvement of memory and body weight *etc*).

There are many other effects of calorie restriction, and so if any compounds are able to mimic some of these effects, then these can be considered as Calorie Restriction Mimetics (CRM). In a paper I wrote some years ago, I suggested that if a compound can mimic at least two biochemical plus one genetic, or five biochemical/physical effects of calorie restriction, then this compound is classified as a CRM [46].

CRM are well-known and some are relatively well-accepted agents which imitate the effects of calorie restriction itself. It is important to highlight the fact that these compounds are not pills for those who are too lazy to diet, but can be used in association with a healthy lifestyle to experience certain benefits which are generally associated with the practice of calorie restriction. It is well-known that calorie restriction can have a host of benefits and has been shown to extend lifespans in several types of animals in the laboratory [47, 48].

The following are some principal CRM agents:

Resveratrol

The action of resveratrol in activating the SIRT family of genes (which mediate some of the effects of CR) is well-documented [49]. As a result of this action, there is an overall reduction of the risk of age-related degeneration.

In a representative study, Marchal *et al.* [50] compared the effects of calorie restriction to the effects of resveratrol, in lemurs. Both interventions reduced oxidative damage in the long term. Interestingly, both interventions increased oxidative stress in the very short term, which is an indication that both act through a hormetic effect, stimulating the organism to execute appropriate repair sequences. This is a typical example of how CRM and other agents can act in order to initiate the beneficial process: by causing mild cellular damage, just enough to arouse the repair process. Resveratrol is an inhibitor of mTOR, and it has been used in association with rapamycin where it was found to have anti-cancer effects [51].

Oxaloacetate

Oxaloacetate increases energy production by the mitochondria, regulates fat tissue production and improves endurance. In a study in mice, it was shown that administration of oxaloacetate for two weeks had an effect on mTOR regulation, reduced inflammatory markers and improved brain metabolism. It also activated the insulin signalling pathway and improved formation of new neurons [52]. Many of these characteristics are also shared by calorie restriction itself, indicating that oxaloacetate can be considered as a typical CRM, and thus, as an effective hormetin.

Oxaloacetate functions by modulating several age-relevant genes and factors, which are also modulated by calorie restriction. The most important of these are the FOXO/DAF-16 factors, and AMPA protein kinase, which is involved in energy production [53]. The authors of this study conclude:

...(Our) results demonstrate that supplementation of the citric acid cycle metabolite, oxaloacetate, influences a longevity pathway, and suggest a tractable means of introducing the health-related benefits of dietary restriction.

Metformin

The actions of metformin as a hormetin have been mentioned above. Metformin also shares boundaries with calorie restriction effects. It is a receptor sensitizer used in diabetes, and it modulates expression of redox protective genes among other actions [54]. Studies show that metformin regulates the insulin/IGF-1 signalling pathway which is one of the most important pathways during ageing [55]. Calorie restriction itself is a modulator of this pathway.

Metformin has several potential anti-ageing actions and it is now being studied in the context of a comprehensive trial as a so-called 'anti-ageing pill' (Targeting Aging with Metformin, or TAME trial [56].

Deoxyglucose

This is one of the first CRM described over 17 years ago. It inhibits glycolysis and mimics some of the effects of CR, particularly increased insulin sensitivity, reduced glucose levels and other biochemical changes. Although using high doses of deoxyglucose can be toxic, research shows that low doses are effective and it is still used in experiments [57].

There are several other potential CRM (Table **1**) which cover a wide spectrum of stress response and metabolic domains.

Table 1 . A list of some candidate CRM and their potential actions.

Compound	Description
aminoguanidine and carnosine	Antiglycating agents
4-phenylbutyrate (PBA)	Histone deacetylase inhibitor, chemical chaperone
Naloxone	Opioid antagonist
Gugulipids	Plant extracts

(Table 1) contd.....

Compound	Description
Olbetam and Rimonabant	Anti-obesity drugs
DPP-4 inhibitors	Oral hypoglycaemics, such as sitagliptin
Iodoacetate	A peptidase inhibitor, hormetic agent
Peptide PYY3-36	Appetite regulator
Rapamycin, and related agents	mTOR inhibitors
Gymnemoside	Triterpenoidal glycoside, anti-obesity pro-drug
Adiponectin	Fatty acid and glucose modulator
Aldifen (2,4-Dinitrophenol)	Hormetic metabolic poison
Galanin antagonists	Neuro-regulators
Modulators of Neuropeptide Y	Can reduce food consumption and appetite
Exanatide	Anti-diabetic
Hydroxycitrate	Modulator of lipid metabolism

Calorie Restriction Mimetics: Conclusions

Research over the past several decades confirms that calorie restriction acts as a hormetic nutritional challenge and can extend health span. However, when it comes to pharmacological interventions it is quite impossible for any single CRM to imitate every single benefit of calorie restriction. This means that CRM need to be used in combination in order to be effective in a wider sense. While many CRM remain experimental, some are available in everyday clinical use, under medical supervision. Examples include metformin, oxaloacetate, resveratrol, carnosine, aminoguanidine and some of the mTOR modulators. These agents modify biochemical and physical parameters which are beneficial, overall, in reducing the impact of certain age-related conditions. However, the importance of considering any therapeutic interventions in a wider and more comprehensive context (including appropriate integration of other hormetic methodologies) is paramount.

The Hormetic Actions of Exercise Mimetics

We can now examine another example of how chemical compounds may activate stress response pathways, *i.e.* act as hormetic stimulants. This is the example of exercise mimetics. Research into calorie restriction mimetics in the manner discussed above, has inspired interest in other areas, where oral drugs or compounds may be able to exhibit the same biological and physiological benefits as the physical action itself. Here, by 'action' I mean any physical activity, not only diet, but also exercise, and even sexual acts. In the previous section I discussed calorie restriction mimetics in some detail. Here I will concentrate specifically on compounds which stimulate the body to act as if it was stimulated by the physical challenge of exercise. The benefits of these compounds are not only relevant to those who may wish to maximise their physical performance but also for people who may be weak, recovering from an operation or have age-related muscle wasting. Exercise mimetics should not be seen as an alternative to exercise but as pharmacological alternatives for people who cannot actually exercise due to illness.

During physical exercise there is activation of molecules and genes such AMPK (adenosine monophosphate-activated protein kinase), SIRT1 protein, peroxisome proliferator-activated receptor gamma coactivator-1alpha (PGC-1α) and peroxisome proliferator activator receptor delta (PPARδ). These function in a coordinated manner to help muscle tissue remodelling [58]. Any agents that can modify the action of these genes or molecules are therefore considered as exercise mimetics.

AICAR

The most publicly-known exercise mimetic is growth hormone (see below), which acts as an anabolic compound that, among other actions, improves muscle mass and strength. But a well-studied and little known exercise mimetic is the compound Acadesine, also known as AICAR (5-amino-1-β-D-ribofura-osyl-imidazole-4-carboxamide) or Amino-Imidazole Carbox-Amide Ribon-ucleotide (Image **1**). This is an AMPK agonist which interferes with the PPARbeta/delta gene which is normally activated during physical exercise, as mentioned above. In one study, researchers who used AICAR in mice have reported that [59]:

> *".....Unexpectedly, even in sedentary mice, 4 weeks of AICAR treatment alone induced metabolic genes and enhanced running*

*endurance by 44%. These results demonstrate that AMPK-PPARdelta
pathway can be targeted by orally active drugs to enhance training
adaptation or even to increase endurance without exercise....*"

To reiterate, exercise mimetics should not be seen as a 'workout pill'. These
should not be thought as substituting the physical challenge of exercise with a pill
which carries similar physiological actions, but as an adjuvant therapy for people
who cannot exercise. However, some people have also been using exercise
mimetics in order to enhance endurance and physical performance. The World
Anti-Doping Agency has prohibited the use of AICAR by competing athletes, as
it is thought to provide an unfair advantage to the user [60]. One possible
drawback to have in mind is that activation of AMPK may worsen the pro-ageing
actions of UV light on the skin, with increased production of ROS and free
radical-induced apoptosis.

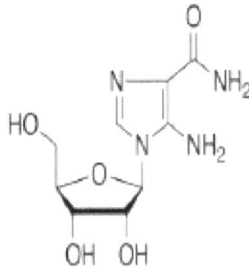

Image. (1). The AICAR Molecule.

Erythropoietin

Production of red blood cells is stimulated by erythropoietin. This agent
stimulates precursors of red blood cells (such as the pro-erythroblasts) found in
the bone marrow, and encourages these to develop into full-grown and active red
blood cells, for optimizing oxygen carrying capacity. In this respect,
erythropoietin has actions similar to exercise which too affects red blood cells and
oxygen-carrying capacity. In addition, erythropoietin stimulates angiogenesis and
promotes the function of smooth muscle fibres. In laboratory animals, treatment
with erythropoietin was found to protect against brain hypoxia, and it also
improves memory and mood (both of which are also improved by physical

exercise itself). When skeletal muscle contracts it produces myokines such as IL-6, BDNF and Irisin which may then affect distant organs such as the brain (and hence the biological basis of exercise positively affecting brain function) [61].

Interestingly, researchers have suggested that physical exercise itself may be considered as a drug, which needs to be administered at an appropriate dose, frequency and duration, with contraindications and side effects [62]. The analogy helps endorse the suggestion that chemical compounds can be used instead of the physical action, based on the same physiological principle of stress response activation.

Lactate

Lactate can enhance certain genes related to exercise, as well as modify other factors such as RNA expression, not only in the muscles or lungs, but also in the liver and brain. Experiments with mice show that Brain TNF (Tumour Necrosis Factor) fell (with a consequent reduced risk of brain inflammation) and vascular endothelial growth factor increased (which improved the function of the blood vessels) after administration of lactate [63]. The authors concluded:

> " ...*exogenously administered lactate was found to reproduce some but not all of these observed liver and brain changes. Our data suggest that lactate, an exercise by-product, could mediate some of the effects exercise has on the liver and the brain, and that lactate itself can act as a partial exercise mimetic.*"

Growth Hormone

The health benefits and drawbacks of growth hormone (GH) are well-known (Image **2**). Here I will concentrate specifically on GH secretagogues, *i.e.* substances which enhance the secretion of GH, and thus are expected to have the same benefits as GH itself, with minimal side effects. One such peptide is the Growth Hormone Releasing Hexapeptide (GHRP6), a synthetic compound which stimulates GH secretion and activity [64]. Another effective secretagogue is GHRP2, which, among other benefits, may gently stimulate the appetite (just like exercise does). These and other GH releasing peptides (GHRPs) act on specific receptors in the brain [65] and may help reduce fat while, at the same time, build-up muscle tissue and reduce insulin resistance if used at the appropriate dose. Athletes have also reported that they experience an increase in endurance and strength. It is clear that these and similar GHRPs belong to the group of exercise mimetics.

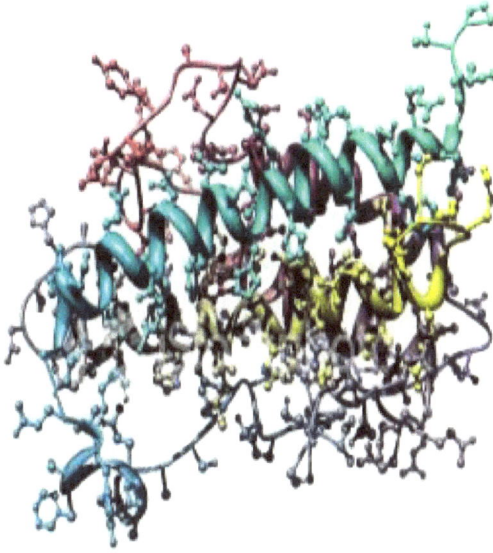

Image. (2). A 3-D structure of the growth hormone molecule.

Some other exercise mimetics

* Carnosine is a well-known booster of muscle activity among its many other benefits [66]. It is mentioned here to highlight the point made earlier about the pluripotent effects of some hormetins.

* Free Fatty Acids (FFA). In an experiment [67], the use of FFA in mice showed a possible role in enhancing other exercise mimetics such as AICAR, and also some physiological changes shared with exercise such as regulation of interleukin factors, and muscle-adipose tissue crosstalk.

* Anabolic steroids. It is possible to include anabolic steroids and their derivatives, such as tetrahydrogestrinone and trenbolone, in the class of exercise

mimetics, in the sense that these hormones affect muscle mass, appetite and performance. However, due to the fact that many are on the banned list and have serious side effects, I will not discuss these further, as their clinical role becomes irrelevant.

* Resveratrol can be considered as a possible exercise mimetic, which has an effect on enhancing endurance [68]. This is thought to be due to activation of the AMPK-SIRT1-PGC-1α pathway.

* Myostatin inhibitors. Myostatin is a protein that, in humans, inhibits muscle differentiation and growth, so any compounds which inhibit its actions would also exert a positive influence on muscle growth and function. Myostatin inhibitors are not currently available for human use but have been used with some success in animals [69].

* Sestrins. These have several effects on metabolism including inhibition of mTOR, antioxidant actions and AMPK-activation [70]. This is particularly relevant both in exercise and in ageing where some changes are shared with those found in non-exercising individuals.

* Arachidonic acid. The enzyme cyclo-oxygenase may metabolise arachidonic acid to prostaglandins which may be considered as anabolic compounds that improve metabolism [71]. The action of these prostaglandins may be blocked by non-steroidal drugs such as ibuprofen. Athletes using arachidonic acid as an oral supplement report increased muscle discomfort after exercise, but at the same time, improved endurance and performance [72].

Exercise Mimetics: Conclusion

Research into exercise mimetics is intense, fuelled by the race to formulate an all-round, effective agent that can mimic as many physiological effects of exercises as possible. Many agents discussed here, as well as other still experimental ones, exhibit a multi-action profile, with pluripotent effects. Exercise mimetics may prove fruitful for weak older patients, those who recover from serious conditions or others who may want to experience a general health-boosting effect. Based on the principle of hormesis, the best result may be achieved by using a combination of different agents, in suitably adjusted dosages.

Sex Mimetics

For the sake of completeness, I will now discuss examples of certain compounds which may mimic some physiological aspects of sexuality (the human ability to have erotic experiences and responses). My intention is to show that hormetic stimulation is a wide-ranging phenomenon, encompassing many domains, including the sexual. The study of these 'sex-mimetics' implies that there are certain common underlying adaptive physiological pathways involving the stress response, a response that can be mimicked by compounds other than those involved in the process of adaptation itself. The result would be that our armamentarium for fighting age-related degeneration based on adaptive dysfunction, becomes wider in scope and more effective in clinical practice. These 'sex-mimetics' are a pharmacological extension of the hormetic sexual techniques discussed in Chapter 1, and as a concept, are complementary to these techniques. This does not necessarily mean that these pharmacological agents are safe and useful for the public at large, but the combination of drug treatment and sexual behavioural techniques may be enough to elicit hormetic/adaptive responses in some individuals, under medical supervision.

Sex Mimetic 1: Flibanserin

A relatively effective treatment of Hypoactive Sexual Desire Disorder (HSDD) is with the multifunctional agonist of the serotonin 5-HT1A receptor, Flibanserin. This is approved specifically for use in HSDD, which is the most common form of female sexual dysfunction. It increases the number of satisfying sexual events in pre and post- menopausal women [73]. Flibanserin is a hormetic agent. Its functions exhibit a 'U-shaped', dose-response pattern: its clinical and pharmacological effects depend on the dose used in each situation [74, 75].

Sex Mimetic 2. Anandamide

The endocannabinoid anandamide is a modulator of neurotransmitter release, which can improve sexual behaviour in animals [76]. It is known that the endocannabinoid system plays a role in regulating sexual behaviour. Anandamide can be seen as a typical 'sex-mimetic' which increases mounting episodes, intromission, ejaculation and resumption of copulation following ejaculation [77]. We know that there is cross-talk between the endocannabinoid and the immune systems [78] where this multifaceted relationship may affect inflammation response, depression and immunocompetence. This underlies the complex relationship that exists between the different systems and hints that there could be ways to modulate one system by regulating the function of another. For instance, the endocannabinoid system regulates neural, endocrine and behavioral responses to stress [79] whereas the immune system is susceptible to the extreme ranges of stressful stimuli. In this respect, modulators of the endocannabinoid systems such as anandamide may be useful in also improving certain aspects of the immune system.

Sex Mimetic 3. Dapoxetine

This is a selective serotonin reuptake inhibitor and antidepressant, used in the treatment of premature ejaculation [80]. While low doses of dapoxetine are clinically useful, higher dosages are detrimental. Rats treated with a high dose of dapoxetine showed a significant increase of sperm abnormalities and sperm motility. This highlights the hormetic, dose-response action of dapoxetine, and reminds us once more of the importance of regulating the degree of challenging stimulation [81].

Sex Mimetic 4. Chlorophytum borivilianum

Commonly known as 'safed musli' (Image **3**), the tuberous roots of this plant have been used not only to improve sexual health but also as an adaptogen (it reduces cellular sensitivity to stress and regulates the stress response), and acts as an immunomodulatory compound. This aligns well with the relationship discussed above, between the stress response, adaptation and modulation of the immune system (as in the case of anandamide). Studies show improvement in sexual function in animals treated with lyophilized aqueous extracts from the roots of *C. borivilianum* [82, 83]. In addition to the sexual element, *C. borivilianum* was found to modulate the inflammatory response, maintain blood glucose, regulate lipid levels and prevent oxidative stress [84, 85].

Image. (3). The plant *Chlorophytum borivilianum.*

The case of 'sex mimetics' is a typical example where information carried by chemical compounds may exert effects at multiple levels and result in different clinical results. Low doses of these compounds is perceived as a challenge by the organism which may result in up-regulation of biochemical pathways, improvement of cross-talk between different biological entities and clinical effects which transcend the boundaries of individual organs or tissues. In fact, the hormetic response can transcend not only organs or tissues but also species: hormetins created by one species in response to stress can be used by another species to adapt to this stress. Here we enter the little-studied and somewhat speculative realm of xenohormesis.

Interspecies Stress Response: Xenohormesis

The logic behind xenohormesis is simple: When plants are under stress they produce different compounds such as polyphenols, which can help them adapt to the effects of the stressful event. When another species (such as humans) consume these chemicals, then the humans are subjected to the protective benefit of the compound, without actually having been exposed to the stressor in the first place. In this way, the hormetic influence has been transferred from the plant to the human. This is the phenomenon of xenohormesis (hormesis initiated by a different species), also described as an 'interspecies epigenetic stress response' (Fig. **2**). More formally, xenohormesis can be described as [86]:

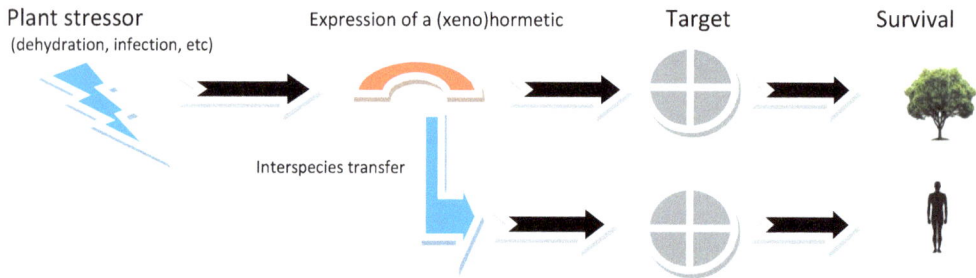

Fig. (2). The pathways of xenohormesis. A plant stressor such as malnutrition, infection or dehydration induces expression of response molecules (xenohormetics) which up-regulate the target molecules in plants and improve the plant's survival. But through interspecies transfer the xenohormetic molecule may also be ingested by a human and it then up-regulates human target domains, thus improve the human stress response, without the human having any previous contact with the stressful stimulus itself.

"... a biological principle that explains how environmentally stressed plants produce bioactive compounds that can confer stress resistance and survival benefits to animals that consume them."

In effect, animals and humans can 'piggyback' off bioactive plant compounds, compounds which reflect the plants sophisticated stress response following millions of years of a stationary lifestyle. The result could be that these xenohormetics can improve human health and function, and have been found to play a part in age-related signalling pathways.

One typical example of a xenohormetic agent is the natural phenol resveratrol which has been described above. Some other xenohormetins are artemisinin, podophyllom, salicylate, curcumin, carotenoids, ascorbic acid, and betalains. Within the concept of xenohormesis, other hormetins, mimetics and agents I discussed above, can be considered xenohormetic agents due to the fact that they originate from plants. An interesting example is the discovery that lithocholic acid (a bile acid) can extend the lifespan of yeast. Because yeast does not synthesize this acid, (but can benefit from it) it has been suggested that this is an example of interspecies hormesis which can be shared by other animals [87].

Many hormetic and longevity pathways are shared between plans and animals. For example, both hormetic and longevity pathways can express heat shock proteins, but there are several other mechanisms. During xenohormesis, animals (including humans) are able to sense (and be challenged by) molecules which were synthesized by plants following a stressful event. In this way, animals and humans can be warned about a forthcoming adverse environment and prepare to react to any future stressful events before these have happened yet. We have to keep in mind however that technology is rapidly changing this equilibrium through the increased use of commercially-grown crops which are either genetically modified or treated with pesticides and other chemicals. This has an impact on the xenohormetic pathway [88] which may be challenging other aspects of our metabolism in order to adapt to the change. In this way we see how technology is affecting our biology, and how it is important to adapt to these changes. Howitz and Sinclair [89] quote:

We propose that the common ancestor of plants and animals synthesized polyphenols. Since the divergence of phyla, there has been selection such that heterotrophs (animals and fungi) detect chemical cues about their environment from plants and other autotrophs (that is, organisms that derive energy from light or inorganic chemical reactions). These chemical cues would give the heterotroph advance warning about the deterioration of the environment, allowing them to prepare while conditions are still relatively favorable. The theory predicts that many key mammalian enzymes and receptors will have evolved binding pockets that allow modulation by molecules produced by other species.

Polyphenols present in olive oil are xenohormetic compounds which exhibit anti-ageing activities [90]. Specifically, the secoiridoid polyphenols from a Mediterranean-style diet containing olive oil, interact with the AMPK/mTOR-axis and exhibit anticancer effects through activation of stress response pathways. Examples include the activation of the endoplasmic reticulum stress response, the unfolded protein response, and an improved sirtuin-1 and NRF2 signalling [91]. These xenohormetics also supress the germ line-like repair processes of cancer cells (*i.e.* they supress the biological immortality and self-renewal capacity of cancer stem cells). The authors quote:

> *...secoiridoids prevent age-related changes in the cell size, morphological heterogeneity, arrayed cell arrangement and senescence-associated β-galactosidase staining of normal diploid human fibroblasts at the end of their proliferative lifespans. EVOO secoiridoids, which provide an effective defense against plant attack by herbivores and pathogens, are bona fide xenohormetins that are able to activate the gerosuppressor AMPK and trigger numerous resveratrol-like anti-aging transcriptomic signatures. As such, EVOO secoiridoids constitute a new family of plant-produced gerosuppressant agents that molecularly "repair" the aimless (and harmful) AMPK/mTOR-driven quasi-program that leads to aging and aging-related diseases, including cancer.*

Another xenohormetic compound is ginseng which has been studied in relation to traditional Chinese medicine [92]. The stress-response effects of ginseng may help align traditional Chinese medicine principles with those of xenohormesis, a fact that underlines the importance of working closely with natural principles in order to improve our own health. Xenohormetic compounds have pleiotropic effects and can modulate several aspects of our metabolism. For instance xenohormetics such as plant polyphenols may target diverse pathways involving stress response, epigenetic regulation, inflammation and oxidation, energy-saving pathways and membrane-dependant regulation [93].

Studying the mechanisms involved in xenohormesis shows that the stress responses following exposure to a plant hormetin are similar to those seen during any exposure to any stressful stimulation, including heat shock [94]. Xenohormetics cause non-specific protein modifications resulting to mild stress and this lead to the activation of adaptation mechanisms such as up-regulation of heat shock protein expression and cellular resistance to stress [95]. The advantage of studying xenohormesis in some detail is that it provides useful insights into the general mechanisms of hormesis and helps us understand better the relationship between stimulus, response and adaptation. Yun and Doux [96] have put forward an interesting hypothesis, suggesting that our taste preferences have evolved based on xenohormetic events. The quote:

*We propose that taste preferences evolved to serve a secondary function--that of xenohormesis. Stress causes organisms to convert complex sugars to simple sugars, as seen during fruit ripening, and to increase the proportion of high-energy saturated fats relative to unsaturated fats, as seen among farmed livestock. **The presence of dietary simple sugars, saturated fats, and salt within an organism may echo its stress experience--an experience assimilated by others when consumed**. As each successive consumer in the food chain incorporates the stress phenotypes of its dietary components, cues for stress may accumulate in a game of "you-are-what-you-eat". Detection of environmental stress embedded in diet may promote adaptive phenotype remodelling such as caloric hoarding to contend with potential ecologic challenges. The phenotype remodelling may be the result of direct stress signalling properties of fats, sugars, and salt. Since food ecosystems typically exhibit seasonality in composition, early detection of cues of ecologic stress during autumn, such as dehydration, lowered ambient temperatures, and impending resource scarcity, likely confers advantages in fitness. Taste preferences may represent a form of "Darwinian rubbernecking." Much like paying attention to vignettes of violence and trauma, recognizing proxies of ecologic stress and adapting accordingly may yield fitness advantages. **Many aspects of agricultural modernization may increase the level of stress embedded in the food chain, catering to pre-existing taste preferences in a form of illegitimate signalling.** Globalization and technology have transformed the dietary experience of autumn--when the food chain undergoes stress and therefore tastes the best--into a year-round bacchanal. Instead of experiencing ecologic stress through their diet in a seasonal pattern, modern humans have become creatures of chronic stress. Many human conditions related to stress dysfunctions may partly arise from maladaptive consumption of stressed foods. We*

anticipate that low-stress and stress-free food may have therapeutic
potential in the treatment of diseases and the promotion of health
(emphasis mine).

These comments reflect the difficulties encountered when technology changes established natural patterns, and may result in excessive and thus unhealthy stimulations. Technological interference with animal rearing and agriculture may cause pathological stress signalling which, through xenohormesis, may be transmitted to humans and result in chronic human stress conditions such as obesity [97]. Modern artificial ingredients of our food or drinks may act as illegitimate signals and mimic stress-response factors, affecting our own metabolism [98]. Nevertheless, it is interesting to speculate that definite stress-related signals are responsible for shaping our own behaviour and health status. And we can also further speculate that despite the changing technological landscape which, initially, may adversely affect our health, our adaptation mechanisms must continue their remodelling, and adjust to the new and 'unnatural' challenges we now encounter. This may well result in increasing robustness and improving our own evolutionary priorities, with reduced likelihood of biological fragility, disease and dysfunction. One interesting experiment at this point would be to study mice fed on a normal (industry-produced) food *vs* organic seasonal food, and examine their overall longevity. According to the argument presented in this chapter, it would be expected that the second group (fed on organic, natural, free from artificial chemicals food) would live longer and be healthier for longer.

CONCLUSION

There is no doubt that there are many pathways involved in the stress response, and that there also exist chemical compounds which can modulate different aspects of this response. The universal nature of hormesis ensures that we can benefit from the action of a wide variety of information-carrying molecules, information which challenges our organism and up-regulates our repair processes. In the discussion above, I made clear that this hormetic adaptation may take place by physically challenging the organism through nutrition, physicochemical actions or exercise, **and/or** through chemical challenges such as mimetics and xenohormetics. Although many details of these processes still remain unclear, it is safe to assume that the general principle of challenging the organism through suitable interventions may well result in acquiring several health assets and can help delay certain aspects of ageing. Modern, technologically-based interventions on our food chain may be a health hazard, but can nevertheless act as a new type of stimulus, inciting action from our adaptive processes. This may ultimately result as increased robustness and reduced likelihood of age-related degeneration.

REFERENCES

[1]   Rattan SI. Hormesis in aging. Ageing Res Rev 2008; 7(1): 63-78.
       [PMID: 17964227]

[2]   Rattan SI, Demirovic D. Hormesis can and does work in humans. Dose Response 2009; 8(1): 58-63.
       [PMID: 20221290]

[3]   Kyriazis M. Nonlinear stimulation and hormesis in human aging: practical examples and action mechanisms. Rejuvenation Res 2010; 13(4): 445-52.
       [PMID: 20662589]

[4]   Rattan SI, Kryzch V, Schnebert S, Perrier E, Nizard C. Hormesis-based anti-aging products: a case study of a novel cosmetic. Dose Response 2013; 11(1): 99-108.
       [PMID: 23548988]

[5]   Rattan SI, Deva T. Testing the hormetic nature of homeopathic interventions through stress response pathways. Hum Exp Toxicol 2010; 29(7): 551-4.
       [PMID: 20558605]

[6]   Rattan SI. Targeting the age-related occurrence, removal, and accumulation of molecular damage by hormesis. Ann N Y Acad Sci 2010; 1197: 28-32.
       [PMID: 20536829]

[7]   Rattan SI. Rationale and methods of discovering hormetins as drugs for healthy ageing. Expert Opin Drug Discov 2012; 7(5): 439-48.
       [PMID: 22509769]

[8]   Calabrese V, Scapagnini G, Davinelli S, *et al*. Sex hormonal regulation and hormesis in aging and longevity: role of vitagenes. J Cell Commun Signal 2014; 8(4): 369-84.
       [PMID: 25381162]

[9]   Cornelius C, Graziano A, Calabrese EJ, Calabrese V. Hormesis and vitagenes in aging and longevity: mitochondrial control and hormonal regulation. Horm Mol Biol Clin Investig 2013; 16(2): 73-89.
       [PMID: 25436749]

[10] Calabrese V, Cornelius C, Dinkova-Kostova A, *et al*. Cellular stress responses, hormetic phytochemicals and vitagenes in aging and longevity. Biochimica et Biophysica Acta (BBA) -. Molecular Basis of Disease 2012; 1822(5): 753-83.

[11] Cornelius C, Koverech G, Crupi R, *et al.* Osteoporosis and alzheimer pathology: Role of cellular stress response and hormetic redox signaling in aging and bone remodeling. Front Pharmacol 2014; 5: 120. [http://dx.doi.org/10.3389/fphar.2014.00120] [PMID: 24959146]

[12] Vasto S, Buscemi S, Barera A, Di Carlo M, Accardi G, Caruso C. Mediterranean diet and healthy ageing: a Sicilian perspective. Gerontology 2014; 60(6): 508-18. [PMID: 25170545]

[13] Krug S, Kastenmüller G, Stückler F, *et al.* The dynamic range of the human metabolome revealed by challenges. FASEB J 2012; 26(6): 2607-19. [PMID: 22426117]

[14] Calabrese EJ. Hormesis and medicine. Br J Clin Pharmacol 2008; 66(5): 594-617. [PMID: 18662293]

[15] Mattson MP, Calabrese EJ, Eds. Hormesis: A Revolution in Biology, Toxicology and Medicine. New York: Humana Press, Springer 2010.

[16] Anisimov VN, Popovich IG, Zabezhinski MA, *et al.* Sex differences in aging, life span and spontaneous tumorigenesis in 129/Sv mice neonatally exposed to metformin. Cell Cycle 2015; 14(1): 46-55. [PMID: 25483062]

[17] De Haes W, Lotte Frooninck L, Van Assche R, *et al.* Metformin promotes lifespan through mitohormesis *via* the peroxiredoxin PRDX-2. Available from: http://www.pnas.org/content/111/24/E2501.full.pdf 2014.

[18] Martin-Castillo B, Vazquez-Martin A, Oliveras-Ferraros C, Menendez JA. Metformin and cancer: doses, mechanisms and the dandelion and hormetic phenomena. Cell Cycle 2010; 9(6): 1057-64. [PMID: 20305377]

[19] Rattan SI, Fernandes RA, Demirovic D, Dymek B, Lima CF. Heat stress and hormetin-induced hormesis in human cells: effects on aging, wound healing, angiogenesis, and differentiation. Dose Response 2009; 7(1): 90-103. [PMID: 19343114]

[20] Son TG, Camandola S, Mattson MP. Hormetic dietary phytochemicals. Neuromolecular Med 2008; 10(4): 236-46. [PMID: 18543123]

[21] Lim YS, Kwon SK, Park JH, Cho CG, Park SW, Kim WK. Enhanced mucosal healing with curcumin in animal oral ulcer model. Laryngoscope 2015; ••• [Epub ahead of print]. [http://dx.doi.org/10.1002/lary.25649] [PMID: 26418439]

[22] Demirovic D, Rattan SI. Curcumin induces stress response and hormetically modulates wound healing ability of human skin fibroblasts undergoing ageing *in vitro*. Biogerontology 2011; 12(5): 437-44. [PMID: 21380847]

[23] Boldyrev AA, Aldini G, Derave W. Physiology and pathophysiology of carnosine. Physiol Rev 2013; 93(4): 1803-45. [PMID: 24137022]

[24]   Calabrese V, Cornelius C, Cuzzocrea S, Iavicoli I, Rizzarelli E, Calabrese EJ. Hormesis, cellular stress response and vitagenes as critical determinants in aging and longevity. Mol Aspects Med 2011; 32(4-6): 279-304.
[PMID: 22020114]

[25]   Sonneborn JS. Mimetics of hormetic agents: stress-resistance triggers. Dose Response 2010; 8(1): 97-121.
[PMID: 20221297]

[26]   Hoffman JR, Ostfeld I, Stout JR, Harris RC, Kaplan Z, Cohen H. β-Alanine supplemented diets enhance behavioral resilience to stress exposure in an animal model of PTSD. Amino Acids 2015; 47(6): 1247-57.
[PMID: 25758106]

[27]   Bex T, Chung W, Baguet A, Achten E, Derave W. Exercise training and Beta-alanine-induced muscle carnosine loading. Front Nutr 2015; 2: 13.
[PMID: 25988141]

[28]   Calabrese V, Cornelius C, Dinkova-Kostova AT, Calabrese EJ. Vitagenes, cellular stress response, and acetylcarnitine: relevance to hormesis. Biofactors 2009; 35(2): 146-60.
[PMID: 19449442]

[29]   Wang D, Wu M, Li S, Gao Q, Zeng Q. Artemisinin mimics calorie restriction to extend yeast lifespan *via* a dual-phase mode: a conclusion drawn from global transcriptome profiling. Sci China Life Sci 2015; 58(5): 451-65.
[PMID: 25682392]

[30]   Zhang XG, Li GX, Zhao SS, Xu FL, Wang YH, Wang W. A review of dihydroartemisinin as another gift from traditional Chinese medicine not only for malaria control but also for schistosomiasis control. Parasitol Res 2014; 113(5): 1769-73.
[PMID: 24609234]

[31]   Das SS, Nanda GG, Alone DP. Artemisinin and curcumin inhibit *Drosophila* brain tumor, prolong life span, and restore locomotor activity. IUBMB Life 2014; 66(7): 496-506.
[PMID: 24975030]

[32]   Barciszewski J, Siboska GE, Pedersen BO, Clark BF, Rattan SI. Evidence for the presence of kinetin in DNA and cell extracts. FEBS Lett 1996; 393(2-3): 197-200.
[PMID: 8814289]

[33]   Sharma SP, Kaur J, Rattan SI. Increased longevity of kinetin-fed Zaprionus fruitflies is accompanied by their reduced fecundity and enhanced catalase activity. Biochem Mol Biol Int 1997; 41(5): 869-75.
[PMID: 9137816]

[34]   Berge U, Kristensen P, Rattan SI. Hormetic modulation of differentiation of normal human epidermal keratinocytes undergoing replicative senescence *in vitro*. Exp Gerontol 2008; 43(7): 658-62.
[PMID: 18262743]

[35]   Barone E, Calabrese V, Mancuso C. Ferulic acid and its therapeutic potential as a hormetin for age-related diseases. Biogerontology 2009; 10(2): 97-108.
[PMID: 18651237]

[36] Lishmanov IuB, Naryzhnaia NV, Maslov LN. [Effect of enkephalins on biosynthesis of myocardial proteins during acute exposure to cold]. Vopr Med Khim 1999; 45(3): 227-31.
[PMID: 10432558]

[37] Inuo H, Eguchi S, Yanaga K, *et al.* Protective effects of a hibernation-inducer on hepatocyte injury induced by hypothermic preservation. J Hepatobiliary Pancreat Surg 2007; 14(5): 509-13.
[PMID: 17909722]

[38] Kumar J, Park KC, Awasthi A, Prasad B. Silymarin extends lifespan and reduces proteotoxicity in C. elegans Alzheimer's model. CNS Neurol Disord Drug Targets 2015; 14(2): 295-302.
[PMID: 25613505]

[39] Bongiovanni GA, Soria EA, Eynard AR. Effects of the plant flavonoids silymarin and quercetin on arsenite-induced oxidative stress in CHO-K1 cells. Food Chem Toxicol 2007; 45(6): 971-6.
[PMID: 17240505]

[40] He W, Zhuang Y, Wang L, *et al.* Geranylgeranylacetone attenuates hepatic fibrosis by increasing the expression of heat shock protein 70. Mol Med Rep 2015; 12(4): 4895-900.
[PMID: 26165998]

[41] Sonneborn JS. Mimetics of hormetic agents: stress-resistance triggers. Dose Response 2010; 8(1): 97-121.
[PMID: 20221297]

[42] Rao CV, Rivenson A, Katiwalla M, Kelloff GJ, Reddy BS. Chemopreventive effect of oltipraz during different stages of experimental colon carcinogenesis induced by azoxymethane in male F344 rats. Cancer Res 1993; 53(11): 2502-6.
[PMID: 8495412]

[43] Kim TH, Eom JS, Lee CG, Yang YM, Lee YS, Kim SG. An active metabolite of oltipraz (M2) increases mitochondrial fuel oxidation and inhibits lipogenesis in the liver by dually activating AMPK. Br J Pharmacol 2013; 168(7): 1647-61.
[PMID: 23145499]

[44] Chattopadhyay D, Sen S, Chatterjee R, Roy D, James J, Thirumurugan K. Context- and dose-dependent modulatory effects of naringenin on survival and development of *Drosophila melanogaster*. Biogerontology 2015. [Epub ahead of print].
[PMID: 26520643]

[45] Marzetti E, Lawler JM, Hiona A, Manini T, Seo AY, Leeuwenburgh C. Modulation of age-induced apoptotic signaling and cellular remodeling by exercise and calorie restriction in skeletal muscle. Free Radic Biol Med 2008; 44(2): 160-8.
[PMID: 18191752]

[46] Kyriazis M. Calorie Restriction Mimetics: examples and mode of action. Open Longev Sci 2009; 3: 17-21.

[47] Colman RJ, Beasley TM, Kemnitz JW, Johnson SC, Weindruch R, Anderson RM. Caloric restriction reduces age-related and all-cause mortality in rhesus monkeys. Nat Commun 2014; 5: 3557.
[http://dx.doi.org/10.1038/ncomms4557] [PMID: 24691430]

[48] Heilbronn LK, Ravussin E. Calorie restriction and aging: review of the literature and implications for

studies in humans. Am J Clin Nutr 2003; 78(3): 361-9.
[PMID: 12936916]

[49]   Chen T, Li J, Liu J, *et al*. Activation of SIRT3 by resveratrol ameliorates cardiac fibrosis and improves cardiac function *via* the TGF-β/Smad3 pathway. Am J Physiol Heart Circ Physiol 2015; 308(5): H424-34.
[PMID: 25527776]

[50]   Marchal J, Dal-Pan A, Epelbaum J, *et al*. Calorie restriction and resveratrol supplementation prevent age-related DNA and RNA oxidative damage in a non-human primate. Exp Gerontol 2013; 48(9): 992-1000.
[PMID: 23860387]

[51]   Alayev A, Berger SM, Kramer MY, Schwartz NS, Holz MK. The combination of rapamycin and resveratrol blocks autophagy and induces apoptosis in breast cancer cells. J Cell Biochem 2015; 116(3): 450-7.
[PMID: 25336146]

[52]   Wilkins HM, Harris JL, Carl SM, *et al*. Oxaloacetate activates brain mitochondrial biogenesis, enhances the insulin pathway, reduces inflammation and stimulates neurogenesis. Hum Mol Genet 2014; 23(24): 6528-41.
[PMID: 25027327]

[53]   Williams DS, Cash A, Hamadani L, Diemer T. Oxaloacetate supplementation increases lifespan in *Caenorhabditis elegans* through an AMPK/FOXO-dependent pathway. Aging Cell 2009; 8(6): 765-8.
[PMID: 19793063]

[54]   de Kreutzenberg SV, Ceolotto G, Cattelan A, *et al*. Metformin improves putative longevity effectors in peripheral mononuclear cells from subjects with prediabetes. A randomized controlled trial. Nutr Metab Cardiovasc Dis 2015; 25(7): 686-93.
[PMID: 25921843]

[55]   Anisimov VN, Piskunova TS, Popovich IG, *et al*. Gender differences in metformin effect on aging, life span and spontaneous tumorigenesis in 129/Sv mice. Aging (Albany, NY) 2010; 2(12): 945-58.
[PMID: 21164223]

[56]   Check Hayden E. Anti-ageing pill pushed as bona fide drug. Nature 2015; 522(7556): 265-6.
[http://dx.doi.org/10.1038/522265a] [PMID: 26085249]

[57]   Ingram DK, Roth GS. Calorie restriction mimetics: can you have your cake and eat it, too? Ageing Res Rev 2015; 20: 46-62.
[PMID: 25530568]

[58]   Matsakas A, Narkar VA. Endurance exercise mimetics in skeletal muscle. Curr Sports Med Rep 2010; 9(4): 227-32.
[PMID: 20622541]

[59]   Narkar VA, Downes M, Yu RT, *et al*. AMPK and PPARdelta agonists are exercise mimetics. Cell 2008; 134(3): 405-15.
[PMID: 18674809]

[60]   http://www.usada.org/wp-content/uploads/wada-2015-prohibited-list-en.pdf

[61] Bo H, Jiang N, Zhang ZY, Ji LL. Exercise and health: from evaluation of health-promoting effects of exercise to exploration of exercise mimetics. Sheng Li Ke Xue Jin Zhan 2014; 45(4): 251-6.

[62] Vina J, Borras C, Sanchis-Gomar F, *et al.* Pharmacological properties of physical exercise in the elderly. Curr Pharm Des 2014; 20(18): 3019-29.
[PMID: 24079769]

[63] Lu J, Selfridge JE, Burns JM, Swerdlow RH. Lactate administration reproduces specific brain and liver exercise-related changes. J Neurochem 2013; 127(1): 91-100.
[PMID: 23927032]

[64] Veldhuis JD, Bowers CY. Integrating GHS into the Ghrelin System. Int J Pept 2010.

[65] Saito Y. [Modulations of human growth hormone receptor level on the cell surface]. Kokuritsu Iyakuhin Shokuhin Eisei Kenkyusho Hokoku 1997; (115): 27-39.
[PMID: 9641816]

[66] Sale C, Artioli GG, Gualano B, Saunders B, Hobson RM, Harris RC. Carnosine: from exercise performance to health. Amino Acids 2013; 44(6): 1477-91.
[PMID: 23479117]

[67] Sánchez J, Nozhenko Y, Palou A, Rodríguez AM. Free fatty acid effects on myokine production in combination with exercise mimetics. Mol Nutr Food Res 2013; 57(8): 1456-67.
[PMID: 23650203]

[68] Hart N, Sarga L, Csende Z, *et al.* Resveratrol enhances exercise training responses in rats selectively bred for high running performance. Food Chem Toxicol 2013; 61: 53-9.
[PMID: 23422033]

[69] Whittemore LA, Song K, Li X, *et al.* Inhibition of myostatin in adult mice increases skeletal muscle mass and strength. Biochem Biophys Res Commun 2003; 300(4): 965-71.
[PMID: 12559968]

[70] Parmigiani A, Nourbakhsh A, Ding B, *et al.* Sestrins inhibit mTORC1 kinase activation through the GATOR complex. Cell Reports 2014; 9(4): 1281-91.
[PMID: 25457612]

[71] Roberts MD, Iosia M, Kerksick CM, *et al.* Effects of arachidonic acid supplementation on training adaptations in resistance-trained males. J Int Soc Sports Nutr 2007; 28: 4-21.

[72] Arsić A, Vučić V, Tepšić J, Mazić S, Djelić M, Glibetić M. Altered plasma and erythrocyte phospholipid fatty acid profile in elite female water polo and football players. Appl Physiol Nutr Metab 2012; 37(1): 40-7.
[PMID: 22165902]

[73] Dhanuka I, Simon JA. Flibanserin for the treatment of hypoactive sexual desire disorder in premenopausal women. Expert Opin Pharmacother 2015; 16(16): 2523-9.
[PMID: 26395164]

[74] Stahl SM. Mechanism of action of flibanserin, a multifunctional serotonin agonist and antagonist (MSAA), in hypoactive sexual desire disorder. CNS Spectr 2015; 20(1): 1-6.
[PMID: 25659981]

[75]   Marazziti D, Palego L, Giromella A, *et al.* Region-dependent effects of flibanserin and buspirone on adenylyl cyclase activity in the human brain. Int J Neuropsychopharmacol 2002; 5(2): 131-40.
[PMID: 12135537]

[76]   Canseco-Alba A, Rodríguez-Manzo G. Anandamide transforms noncopulating rats into sexually active animals. J Sex Med 2013; 10(3): 686-93.
[PMID: 22906359]

[77]   Rodríguez-Manzo G, Canseco-Alba A. Anandamide reduces the ejaculatory threshold of sexually sluggish male rats: possible relevance for human lifelong delayed ejaculation disorder. J Sex Med 2015; 12(5): 1128-35.
[PMID: 25808995]

[78]   Boorman E, Zajkowska Z, Ahmed R, Pariante CM, Zunszain PA. Crosstalk between endocannabinoid and immune systems: a potential dysregulation in depression? Psychopharmacology 2015.

[79]   Lee TT, Hill MN, Hillard CJ, Gorzalka BB. Disruption of peri-adolescent endocannabinoid signaling modulates adult neuroendocrine and behavioral responses to stress in male rats. Neuropharmacology 2015; 99: 89-97.
[PMID: 26192544]

[80]   Yue FG, Dong L, Hu TT, Qu XY. Efficacy of Dapoxetine for the treatment of premature ejaculation: a meta-analysis of randomized clinical trials on intravaginal ejaculatory latency time, patient-reported outcomes, and adverse events. Urology 2015; 85(4): 856-61.
[PMID: 25817107]

[81]   ElMazoudy R, AbdelHameed N, ElMasry A. Paternal dapoxetine administration induced deterioration in reproductive performance, fetal outcome, sexual behavior and biochemistry of male rats. Int J Impot Res 2015; 27(6): 206-14. [Epub ahead of print].
[http://dx.doi.org/10.1038/ijir.2015.16] [PMID: 26399566]

[82]   Thakur M, Chauhan NS, Bhargava S, Dixit VK. A comparative study on aphrodisiac activity of some ayurvedic herbs in male albino rats. Arch Sex Behav 2009; 38(6): 1009-15.
[PMID: 19139984]

[83]   Kenjale R, Shah R, Sathaye S. Effects of *Chlorophytum borivilianum* on sexual behaviour and sperm count in male rats. Phytother Res 2008; 22(6): 796-801.
[PMID: 18412148]

[84]   Lande AA, Ambavade SD, Swami US, Adkar PP, Ambavade PD, Waghamare AB. Saponins isolated from roots of *Chlorophytum borivilianum* reduce acute and chronic inflammation and histone deacetylase. J Integr Med 2015; 13(1): 25-33.
[PMID: 25609369]

[85]   Giribabu N, Kumar KE, Rekha SS, Muniandy S, Salleh N. *Chlorophytum borivilianum* root extract maintains near normal blood glucose, insulin and lipid profile levels and prevents oxidative stress in the pancreas of streptozotocin-induced adult male diabetic rats. Int J Med Sci 2014; 11(11): 1172-84.
[PMID: 25249786]

[86]   Hooper PL, Hooper PL, Tytell M, Vígh L. Xenohormesis: health benefits from an eon of plant stress response evolution. Cell Stress Chaperones 2010; 15(6): 761-70.

[PMID: 20524162]

[87]   Goldberg AA, Kyryakov P, Bourque SD, Titorenko VI. Xenohormetic, hormetic and cytostatic selective forces driving longevity at the ecosystemic level. Aging (Albany, NY) 2010; 2(8): 461-70.
       [PMID: 20693605]

[88]   Gregoraszczuk EL, Milczarek K, Wójtowicz AK, Berg V, Skaare JU, Ropstad E. Steroid secretion following exposure of ovarian follicular cells to three different natural mixtures of persistent organic pollutants (POPs). Reprod Toxicol 2008; 25(1): 58-66.
       [PMID: 18024081]

[89]   Howitz KT, Sinclair DA. Xenohormesis: sensing the chemical cues of other species. Cell 2008; 133(3): 387-91.
       [PMID: 18455976]

[90]   Surh YJ. Xenohormesis mechanisms underlying chemopreventive effects of some dietary phytochemicals. Ann N Y Acad Sci 2011; 1229: 1-6.
       [PMID: 21793832]

[91]   Menendez JA, Joven J, Aragonès G, *et al.* Xenohormetic and anti-aging activity of secoiridoid polyphenols present in extra virgin olive oil: a new family of gerosuppressant agents. Cell Cycle 2013; 12(4): 555-78.
       [PMID: 23370395]

[92]   Qi HY, Li L, Yu J. [Xenohormesis: understanding biological effects of traditional Chinese medicine from an evolutionary and ecological perspective]. Zhongguo Zhong Yao Za Zhi 2013; 38(19): 3388-94.
       [PMID: 24422414]

[93]   Barrajón-Catalán E, Herranz-López M, Joven J, *et al.* Molecular promiscuity of plant polyphenols in the management of age-related diseases: far beyond their antioxidant properties. Adv Exp Med Biol 2014; 824: 141-59.
       [PMID: 25038998]

[94]   Lamming DW, Wood JG, Sinclair DA. Small molecules that regulate lifespan: evidence for xenohormesis. Mol Microbiol 2004; 53(4): 1003-9.
       [PMID: 15306006]

[95]   Ohnishi K, Ohkura S, Nakahata E, *et al.* Non-specific protein modifications by a phytochemical induce heat shock response for self-defense. PLoS One 2013; 8(3): e58641.
       [PMID: 23536805]

[96]   Yun AJ, Doux JD. Unhappy meal: how our need to detect stress may have shaped our preferences for taste. Med Hypotheses 2007; 69(4): 746-51.
       [PMID: 17374557]

[97]   Yun AJ, Lee PY, Doux JD. Are we eating more than we think? Illegitimate signaling and xenohormesis as participants in the pathogenesis of obesity. Med Hypotheses 2006; 67(1): 36-40.
       [PMID: 16406352]

[98]   Yun AJ, Doux JD. Stress dysfunctions as a unifying paradigm for illness: repairing relationships instead of individuals as a new gateway for medicine. Med Hypotheses 2007; 68(3): 697-704.

CHAPTER 3

Environmental Enrichment: General Concepts and Research

Abstract: The environment plays a huge role both during ageing and in relation to the phenomenon of hormesis. In this chapter I discuss the impact of any stressful stimuli or challenges originating from our environment, and expand the concept of hormesis to take into account the environment in a wider sense. An enriched environment is taken to mean an 'information-rich' habitat, including the immediate surroundings of an organism. In the case of humans, these surroundings include not only the physical aspects such as towns, natural landscapes and weather, but also social and virtual elements such as online environments and digital relationships. The discussion lays the foundations for understanding how enriched environments act as vehicles of information which lead to biological modifications. These biological modifications may then participate in a novel evolutionary event, which is the emergence of techno-culture, an amalgam of biology and technology. It would be unthinkable to consider human ageing without referring to these new technological developments. Speculative elements such the notion of the 'noeme' (a biological- digital entity), empirical research, emerging research and other concepts are discussed within a mutually-influencing landscape, with the emphasis being on the biology of human ageing

Keywords: Cognition, Epigenetics, Environmental enrichment, Gut microbiota, Indefinite lifespan, Internet, Molecular pathological epidemiology, Natural environment, Noeme, Rejuvenation, r-k model, Social enrichment, Technoculture.

INTRODUCTION

We know that plasticity of the nervous system, and specifically of the synapses, neurogenesis, and general neuronal health, is subjected to external influences, including an intervention which is formally called Environmental Enrichment (EE) [1, 2]. **EE implies being exposed to an information-rich environment**

Marios Kyriazis

which augments the organism's physical and social functions (for example [3]). Such an environment can modulate concentrations of several neuroactive compounds such as glutamate, GABA, dopamine, acetylcholine and others [4].The result of this stimulation is a hormetic effect, whereby mild external challenges (hormetic stresses) up-regulate the function of neural and other elements [5].

There are many studies confirming that EE is a useful strategy in maintaining brain health [6, 7], but research on whether EE has widespread effects on other aspects of physical health is not as widely available. One may argue that a well-functioning brain cannot exist in a badly-functioning body, and thus it can be extrapolated that if EE positively affects the brain it must, somehow, also affect the health of the rest of the body.

What is an Enriched Environment?

An enriched environment is one that consists of substantial amounts of data or information which is meaningful and useful to the organism (*i.e.* non-trivial): - the information entices the organism to respond to the stimulus in order to adapt to it. An information-rich environment may be one that it is richly natural (*i.e.* containing an abundant variety of trees, plants and wildlife [8]). Or it may be a structurally striking city, a stimulating neighbourhood (both socially and architecturally) or an evocative virtual environment (online games or stimuli). Exposure to such an environment affects not only mood, vitality and cognition, but it also provokes distinct biological changes [9] including changes in dopamine and in other neurontransmitters, and modifies several genetic elements. For example, one well-studied gene that has been directly implicated in environmental change is the dopamine receptor gene D4 (DRD4 gene) and its various alleles/polymorphisms. This mediates challenges originating from the environment, and results in distinct phenotypes. It can predict longevity [10] and it is also influenced by the season of birth [11] which supports the notion that the environment may directly affect genetic inheritance.

Natural landscapes may improve other neuronal 'feel-good' factors and result in mental well-being and optimism for the future. This depends on the degree, value and weight of the information carried by the environment [12]. Although in a natural environment one is exposed to a combination of visual, auditory and other stimulation, the health benefits may be experienced even if one is exposed to just one type of stimulation. Pati *et al.* [13] have shown that patients who are exposed to a simulated natural environment (ceiling-mounted photographic sky compositions, in a hospital setting) experience a significant improvement in patient outcomes, with lowering of the blood pressure, a reduced need for medication, reduction of anxiety and stress, and improved environmental satisfaction. This indicates that stimulating even one of the senses can still result in biological/physiological changes as long as the value of information remains relevant, and is able to provoke an adaptive response.

On the other hand, it is also known that modern societies are slowly moving away from nature and espouse a more virtual/artificial setting. The definition of 'natural' is crucial in this discussion. **Natural is an environment which is not made or caused by humans**. This is important because I will later discuss how our environment is becoming more technological and less natural, although one may argue that it is natural for humans to develop (and live in) technological environments.

It was shown that time spent watching television, playing video games or internet use is inversely proportional to hours spent in raw nature [14]. This is characteristic of our modern techno-cultural milieu and it is not necessarily adverse to health, as it is also known that computer use can have health benefits, and it is associated with increased longevity (Fig. **1**).

Being cognitively stimulated in a virtual setting is shown to have several health benefits. For instance, playing online action video games (AVGs) affects the plasticity of sensorimotor regions in the grey matter, and improves connectivity between neurons, particularly those involved in attention and experience [16]. These cognitive benefits of interaction with technology are not specific to action video games, but other active online experiences may also improve cognition. Use of general video games for one hour a day improves spatial memory, cognitive control and complex verbal span [17]. General use of the internet has been studied in a variety of settings and initial results show health promoting effects, including an increase in life expectancy, as discussed in Fig. (**1**) but although there is correlation, the causation is still unclear. A study in Taiwan found that internet use (broadband, wireless or mobile) is positively associated with quality of life, self-esteem and reduction in stress [18]. Another study found positive health improvements following internet use by older people [19].

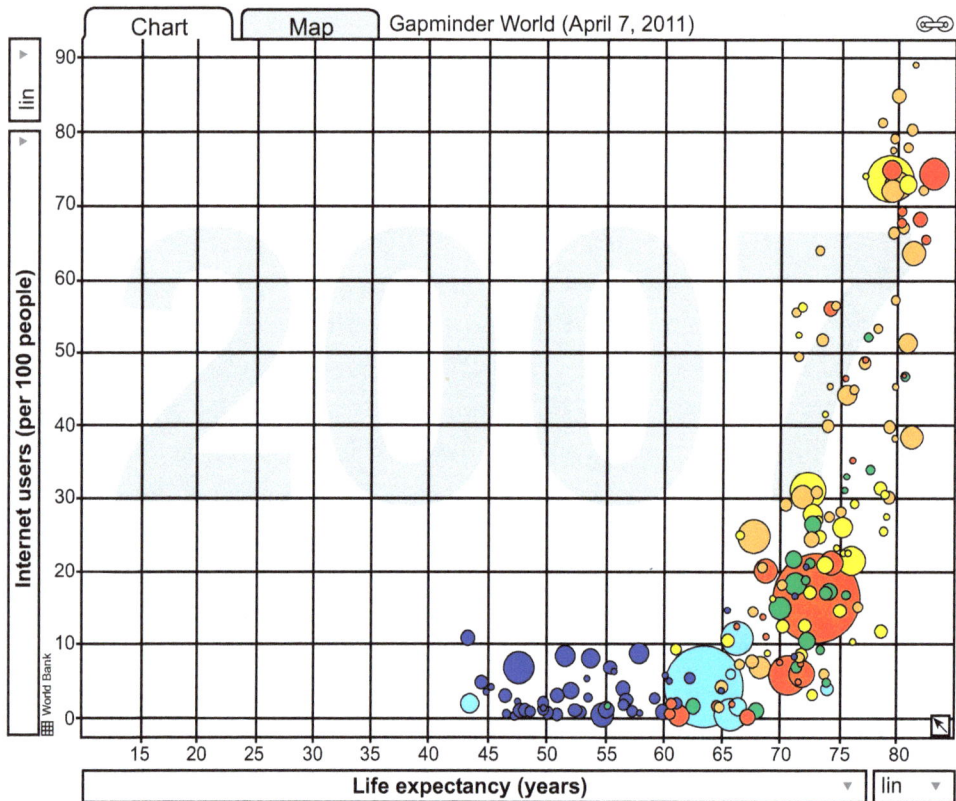

Fig. (1). Correlation between life expectancy and internet users per country. The figure shows an apparent increase in life expectancy as the numbers of internet users increase. Similar correlations can be found between life expectancy and cell phone users (available here: www.bit.ly/1CH2S7l), and also, life expectancy and broadband subscribers per 100 people (available here: www.bit.ly/1CH2LbP). The figure indicates the impact of a modern techno-culture upon life expectancy. However, the causal link between these variables is far from proven. The time scales involved are too short to allow for critical evaluation of any direct causal link between internet use and life expectancy. It is but a preliminary suggestion that increased meaningful use of digital communication technology may influence life expectancy. In addition, it is not merely a matter of how an average internet use can impact on life expectancy, but it is the quality of use (it is qualitative rather than quantitative). It would be necessary to study the impact of 'meaningful' use of internet upon life expectancy, but data for this are not available at present. Image from www.gapminder.org (used under the Creative Commons Attribution 3.0 Unported license) [15].

The 'Noeme' as a Facilitator of Information Exchange

An enriched online environment facilitates exchange of meaningful information and plays a role in shaping our noeme, our virtual and real persona (see [15, 20]). I proposed this term to denote a novel evolutionary replicator, in line with genes and memes. The noeme is the intellectual presence of a physical person within the Global Brain (a self-organising worldwide network of human and information and communication technologies).

While in a virtual domain an avatar is the graphical representation of the user, in the real world a noeme is the abstract representation of one's bio-digital presence and it is identified by the information-sharing strategies of the individual. It involves a merging of the biological with the digital and heralds a period when the line between the physical and the virtual domains becomes increasingly blurred [21, 22]. This blurring, I believe, will catalyse a more favourable situation which can promote extreme healthy human longevity, and I further explain this argument below. In the meantime, one way to enrich one's personal environment is to live in two or more societies simultaneously: one or more real and one or more virtual societies. This is quite achievable. Many thousands of people already live in two real societies simultaneously. For example, people who live in one country for many months and in another country for the rest of the time, switching from one society to another without severing the bonds with the society they leave behind. Moreover, some of these people also belong to virtual societies such as Second Life and/or MMORPGs (Massively Multiplayer Online Role Playing Games). Although there are many studies examining the relationship of addiction and online games, two out of three psychiatrists do not believe that there is a negative association between MMORPG and psychopathology [23].

The biological relevance of living simultaneously in as many societies as possible is this: The increased value of information input reaching the brain, together with the intellectual and emotional stimulation associated with living in such environments activates neural, immune and hormonal pathways which increase the rate of tissue repair. These benefits are essentially based upon up-regulation of neuro-biological elements by a stimulating, enriched environment [24 - 26]. These mechanisms are encompassed by the notion of Hormetic Environmental

Enrichment, whereby an increased stimulation or challenge up-regulates biological repair mechanisms resulting in measurable benefits to the organism, whereas excessive or inadequate stimulation results in malfunction [27]. The basis of these concepts has already been discussed in the previous two chapters.

One may argue that in order for people to be healthy and live longer, their societies (environment) also need to be healthy and live longer, and in order to do this, we need to increase the input of meaningful information into both the person and the society as a whole. Therefore, cognitive activities that increase informational input are useful in this respect. For instance, a virtual society based on MMORPGs (but **not** an addiction to MMORPGs) provides challenging environments with continual neurosensory positive or negative feedback. It engages its 'citizens' to construct a character by performing prestige-gaining actions, and by developing sensory, physical and emotional activities that they would not have performed otherwise [28]. A 'Second Life' environment may be used, for example, as a virtual conference facility, where useful flow of actionable information between its citizens can be facilitated. This augments the information content of the society, heightens its robustness and reliability, and makes it more durable, *i.e.* 'fitter' in evolutionary terms.

One consequence of this is that individuals who live in such societies embody an increasingly indistinguishable combination of their real self and that of their online avatar. Their noeme is being continually enhanced and their information content increases [20]. In order for this process to be more successful however, one has to have avatars and online user names that are identical to their own name in the real world. This can be a pseudonym, but it has to be the same across all platforms. The resulting robustness of information-sharing capabilities strengthens the overall standing of the individual and thus enriches their noeme. I have argued elsewhere that a strong noeme is essential in promoting extreme lifespans, whereas weak noemes are eventually eliminated by evolutionary forces [21]. This is based on the cybernetic concept of Selective Reinforcement, whereby an appropriate agent (or action) is selected and retained if its content of information is of a sufficient and appropriate magnitude [29].

Research has shown that an online avatar can live longer (through user retention) within a society, if this avatar is well connected within that particular society [30] and is retained less if it is not well interacting within that environment [31]. What seems to matter in the retention of an avatar is the number and strength of its connections *i.e.* their 'integration' with the society they live in. This is a universal theme also encountered in the retention (survival) of neurons in the brain [32]. Therefore, taking the above discussion into consideration it is possible to study several domains of EE (natural or virtual):

1. ***Social Enrichment*** *i.e.* physical or virtual interactions with other humans. This can involve a mixture of face to face interactions and virtual/online interactions, such as social media involvement, blogging, *etc*. In this context, information flows freely between agents and this exchange includes (or should include) challenging information that entices us to act in order to force an improvement. This situation is encountered in those who live an active cognitive life not only in a physical sense but also in a virtual environment. Many people choose anonymity while on line. This has no positive effects upon the user's persona as a whole (*i.e.* it does not enhance their noeme) therefore missing an opportunity to strengthen and improve their connections and information-sharing capabilities. A low number of connections, associated with a reduced strength of those connections fails to integrate one's presence in the world, making them easily expendable in the evolutionary sense [33].

   As discussed in chapter 7, the ageing process is demarcated by increasing entropy and loss of useful information input due to a progressively reduced number of challenges. Ageing has been described as a loss of useful biological functional complexity [34]. In order to increase functional complexity and thus increase the information content and reliability of our biological reserves, it is possible to use a basic universal principle which suggests that complexity within a system increases when the variety and connection of individual parts of that system increase with regards to:

   a. space (more numerous)
   b. time (more frequent), and
   c. dynamical scale (more powerful).

To show that complexity has increased, it is necessary to demonstrate that connections have increased in at least one of these dimensions. Using this as a metaphor, in the case of real human avatars (*i.e.* noemes), I suggest that in order to increase complexity, and thus increase their evolutionary fitness in a niche of modern technological environment, one could invoke social challenges by:

- Having a **large number** of connections both in virtual and in real terms. In practice this could mean that in order to be well connected one should, for example, not reject any 'friend' or 'connect' request from anybody, either online or in reality.
- Increasing the **unity** of their connections. Use only one (user)name for all environments across all platforms.
- Increasing the **strength** of their connections. One should provide as much information about themselves as possible, and join as many online information-sharing facilities as practicable.

2. ***Praxeological and Nutritional Enrichment*** by encouraging certain activities that require a physical action. This includes a Paleo lifestyle and exercises, power law exercises, as well as hormetic nutritional exposure such as calorie restriction and intermittent fasting, with irregular meals and ever changing content of the diet. These types of activities have been discussed in chapters 1 and 2. Within this context one may also consider sexual enrichment, introducing variety and novelty in sexual practices [35].

3. ***Habitat-Based Enrichment*** of the surrounding physical environment such as town planning. Takano *et al.* have shown that suitable residential development can positively influence the longevity of older people living in heavily populated megacities [36]. They examined the association between survival and green environment (containing a variety of stimulants such as visual, olphactory and auditory challenges) in Tokyo. They found that the nearer to a green space a resident lived, the more likely it was for this resident to live longer. This and other similar studies highlight the importance of suitable, balanced stress/stimulation. Other stimuli such as music and art resonate with our inner natural self and may excite mechanisms to invoke adaptation. External information such as music directly affects genes and RNA [37]. Kanduri *et al.* [38] have shown that exposure to music (*i.e.* musical information) as a cognitive challenge, influences transcription, improves the

function of the human transciptome, up-regulates dopamine and neuronal protection as well as invoking long term potentiation, and downregulation of components of apoptotic cascade. Translated into clinical terms, this may mean that music-oriented challenges (perhaps activities such as listening to unusual music, foreign songs and computer-generated melodies) can have positive health benefits [39]. However, it is also important to pay attention to the degree and duration of environmental challenges. Based on principles of hormesis mentioned in the previous chapters, it is necessary to balance the stimulus according to a 'dose-response' methodology. Living in a megacity exposes its inhabitants to increased stimulation (from noise, visual stimuli, vibration and olfactory stimuli). These stimuli combine to produce an enhanced and integrated experience, which, if prolonged, may lead to dysfunction and chronic stress [40]. This highlights the importance of being exposed not just to any stimulation but to stimulation that has adaptive relevance, it is novel and not excessive, that requires us to act because it would be better to act than not to act. Continual exposure to information that does not entice us to act in some positive way, will eventually result in annoyance, dissatisfaction, dysfunction, and chronic health conditions [41, 42]. In addition, the issue of habituation to the stimulus need to be addressed. Exposure to the same stimulus for a certain period of time results in habituation and the stimulus ceases to be of any relevance. That is why it is important to be exposed to novel (ever-changing) types of stimuli or challenges. When we develop habituation to a stimulus, then this stimulus ceases being hormetic and becomes either irrelevant or chronically damaging.

4. ***Mental Enrichment*** with internally generated stimuli *i.e.* mindfulness [43]. This includes a wide range of activities and many different practices. Examples may include mindful observation exercises, enhanced and concentrated listening of music, breathing and visualisation, touch awareness and introspection, and oriental-inspired techniques. Some examples are given in the Appendix. Mindfulness stimulation improves emotions, self-compassion, self-esteem, and other neural and behavioural elements [44, 45]. Appropriate mindfulness techniques can also improve other health and social parameters [46]. Mental enrichment exercises are based on increasing the flow of internally-generated information, which has a similar biological and

physiological basis as any externally-derived stimulus. This internally-generated information integrates signals to the amygdala and anterior insula as discussed in chapter 1.

Returning to Basics: Some Mechanisms of EE

Based on animal research we may be in a position to extrapolate and discuss the importance of EE on humans. As we have seen, the environment plays a definite and relevant role in human health and well-being, including social well-being. The fact that the environment can have epigenetic-mediated impact on our phenotype has been addressed by Fraga *et al.* [47]. They reported that monozygotic twins exhibit phenotypic differences due to epigenetic-mediated influence of the environment. They quote:

> *"We found that, although twins are epigenetically indistinguishable during the early years of life, older monozygous twins exhibited remarkable differences in their overall content and genomic distribution of 5-methylcytosine DNA and histone acetylation, affecting their gene-expression portrait. These findings indicate how an appreciation of epigenetics is missing from our understanding of how different phenotypes can be originated from the same genotype".*

This supports the view that an appropriate environment (*i.e.* external information acting internally) will have definite effects on the phenotype. I discussed elsewhere how some of these effects can be mediated by microRNA action, but there are many other mechanisms. One way such a modification can occur is *via* histone modifications. The environment may cause variable expression of metastable epiallele genes and this is influenced by histone modifications. In other words, the environment (through information) can cause epigenetic changes, which can influence metastable epiallele genes (which then cause new phenotypic expression) [48]. More details on metastable epiallele function have been discussed by Dolinoy *et al.* [48]:

> *The phrase "metastable epiallele" was first coined by Rakyan and*

colleagues in 2002 to describe genes that are variably expressed in genetically identical individuals as a result of epigenetic modifications established during early development. The term "epiallele" refers to "an allele that can stably exist in more than one epigenetic state, resulting in different phenotypes," while "metastable" highlights the "labile nature of the epigenetic state of these particular alleles." Metastable epialleles are most often associated with retroelements and transgenesis, resulting in ectopic or aberrant transcription of nearby genes. ...The characterization of histone modifications at metastable epialleles is an important step in identifying mechanisms of action promoting environmentally-induced alterations within the epigenome.

The epigenetic modifications may affect both genetic and personality elements. Lansade *et al.* [49] studied horses subjected to an enriched environment and found that there were significant improvements in behaviour and learning, with some changes persisting for over three months after exposure. In addition:

"..The EE induced the expression of genes involved in cell growth and proliferation, while the control treatment activated genes related to apoptosis."

This is significant and it suggests that an enhanced environment can, in essence, reduce some of the effects of ageing *via* modulation of apoptotic cascades.

Another important effect of EE is that it can up-regulate Brain Derived Neurotrophic Factor (BDNF) an important trophic factor involved in learning and memory (Fig. **2**). Although this factor can also increase after physical exercise, other researchers [50] have not found such a correlation, a fact suggesting the cognitive stimulation is more important than physical stimulation in up-regulating the function of BDNF. This may have relevance in a more profound argument, which concerns the importance of physical exercise in a modern cognition-based society. I have informally argued (in this book and elsewhere) that humans who live in a complex, information-rich environment may experience health benefits which are comparable to those gained from exercise, without the evolutionary need to exercise physically. Certainly, the energy requirements used during cognitive activities can be comparable to the energy requirements encountered during running long distance [51].

Hormesis: a theoretical phenomenon of dose-response relationships in which something (as a heavy metal or ionizing radiation) that produces harmful biological effects at moderate to high doses may produce beneficial effects at low doses

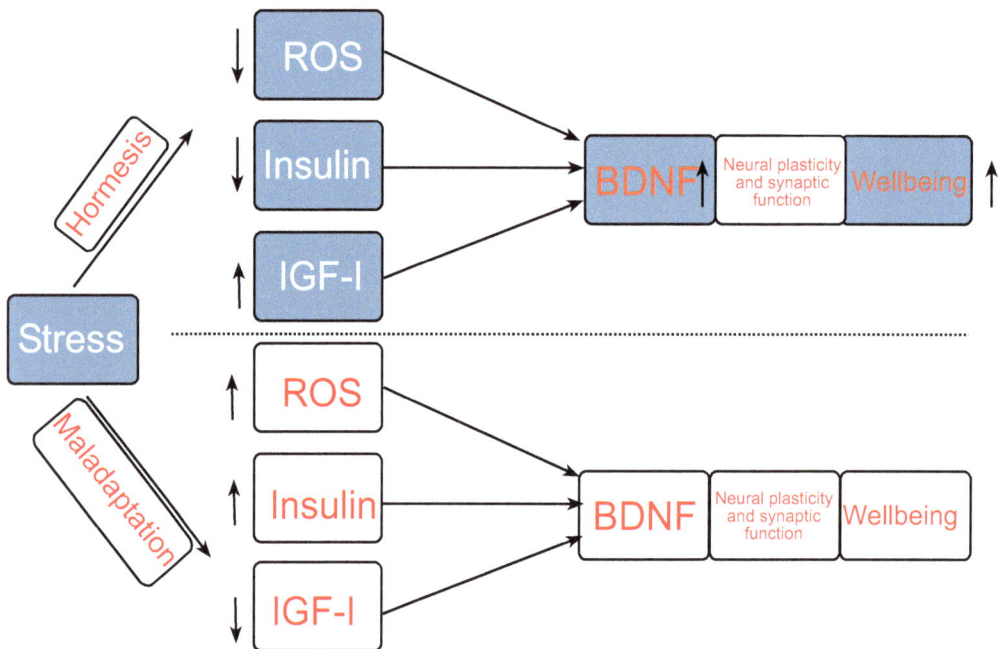

Fig. (2). Diagram showing the effect of environmental stress (positive or negative) on BDNF and neural function. Image from: Professor Grant Schofield. http://profgrant.com/2013/08/06/schofields-hormetic-theory-of-wellbeing/. By kind permission of the author.

This concept is tentatively supported by other work [52] which shows that epigenetic modulation of chromatin, which is needed for transcription control in the hippocampus, is adversely affected by ageing but positively corrected following cognitive enrichment. Cognitive enrichment epigenetically up-regulates the methylation of histone lysine, and as a result, age related decline in memory can be improved, even in the absence of physical exercise. Memory depends on histone methylation in the hippocampus and it is known that environmental influences can affect this process [53]. Modulation of the environment by increasing its information content is perceived as a typical 'challenge' by the hippocampus and it activates epigenetic mechanisms leading to modulation of histone methylation [54]. These authors describe how neurons born from endogenous stem cells subsequently migrate to the dentate gyrus. Adult neurogenesis however, declines with increasing age. EE techniques have been used in order to reverse this decline in neurogenesis, sometimes successfully. The authors conclude [54]:

"Adult hippocampal neurogenesis in mice living in an enriched environment from the age of 10 to 20 months was fivefold higher than in controls. Relatively, the increase in neuronal phenotypes was entirely at the expense of newly generated astrocytes. This cellular plasticity occurred in the context of significant improvements of learning parameters, exploratory behavior, and locomotor activity. Enriched living mice also had a reduced lipofuscin load in the dentate gyrus, indicating decreased nonspecific age-dependent degeneration. Therefore, in mice signs of neuronal aging can be diminished by a sustained active and challenging life, even if this stimulation started only at medium age. Activity exerts not only an acute but also a sustained effect on brain plasticity".

This is a great piece of work, indicating that, in principle, environmental challenges carrying information may be used in order to diminish the impact of ageing, even (as an extrapolation), in humans. The effects of environmental enrichment on ageing parameters depend on the age of the organism.

A discussion about the mechanisms and effects of EE would not be complete without examining how environmental challenges may affect our physiology by acting through the gut interface. Certain mechanisms of EE within the gut provide intriguing insights into the entire concept of EE, and although the discussion below is rather limited as it is not entirely within the scope of the book, it is nevertheless necessary.

The Environment and the Gut Microbiota

The physical presence of gut microbiota is considered here as being part of the external environment which influences and challenges any human organism [55]. Therefore the microbiota play an essential role in shaping the stress response and adaptation capabilities of the host organism, and its actions must be discussed in some basic detail. The gut microbiota can be considered as an interface between the outside environment and our internal milieu. Microbiota can determine and buffer the rate and efficiency of nutrient absorption, thus providing a cushioning mechanism between our external and internal environments [56]. External stimuli and dietary challenges, such as different nutrients, can have definite effects on our biology but these are buffered by the activity of our gut microbiota. For example, a diet high in cholesterol may not necessarily increase cholesterol blood levels, because gut bacteria are able to metabolise cholesterol **before** it enters our blood stream [57]. Bifidobacteria (specifically *Bifidobacterium bifidum* PRL2010) have been shown to be able to assimilate cholesterol and/or convert it to coprostanol, resulting in lower blood cholesterol, at least in animal studies. In humans, intestine microorganisms synthesize large amounts of utilizable folate, biotin and vitamin K even in the absence of these compounds from the diet [58]. It has been estimated [59] that there are over 800 bacterial species and a staggering 7000 different strains in the human colon, approximately 100 trillion microorganisms.

One potential drawback is the fact that the microbiota may also produce toxic by-products such as hydrogen sulfide. However, this may not be as bad as it sounds, because hydrogen sulfide may, if present in low amounts, behave hormetically and result in further benefits to the organism. Hydrogen sulfide (H_2S) is a signalling molecule with a variety of actions, as discussed in previous chapters. Depending on the dose, it may act as a pro-inflammatory or an anti-inflammatory, pro-apoptotic or anti-apoptotic: it is, in other words, a true hormetic compound [60]. It has been suggested that H_2S is at least as important in biology as nitric oxide and carbon monoxide [61]. This indicates that despite the fact that a process can produce an apparently noxious compound, it may be that, on further study, this proves to actually be beneficial, as long as it is based on hormetic principles.

Other research shows that microbiota metabolites can influence the production of serotonin and thus have an impact on mood [62]. For instance, Yano *et al.* [63] have shown that gut microbiota regulate 5-HT and thus serotonin synthesis (as well as platelet function) thus improving feelings of depression. Gut microbiota also affect the immune system and modulate inflammatory markers [64]. The finding that gut microbiota can regulate some brain functions (through the brain-gut axis, and the enteric nervous system) highlights the importance of further study of this environmental interface. For instance, O'Sullivan *et al.* [65] have shown that BDNF activity in the hippocampus can increase following oral probiotics, showing that there is a complex relationship between different environmental stimuli that can result in the same neural effects (in this case BDNF can be increased through exercise [66] and/or probiotics [65] and/or cognitive effort [67]). The gut microbiota has been described as an extension of the brain in the sense that there is communication and cross-talk between the brain and the gut. The human gut microbiome contains genes which outnumber the human gene pool in the rest of the body [68]:

> *Genes within the human gut microbiota, termed the microbiome, significantly outnumber human genes in the body, and are capable of producing a myriad of neuroactive compounds. Gut microbes are part of the unconscious system regulating behavior. Recent investigations indicate that these microbes majorly impact on cognitive function and fundamental behavior patterns, such as social interaction and stress management. In the absence of microbes, underlying neurochemistry is profoundly altered.*

The eye-watering implication of this is that the gut microbiota interface can regulate our behaviour acting in tandem with the central nervous system, to an extend which has not hitherto been thought possible. As an aside, this is a lethal argument against those who expect that technology will one day be able to allow us to upload our brain to a computer: this approach entirely ignores the neuro-behavioural input of the gut microbiota, input which shapes us into our existing self.

We are now in a position to examine the role of the environment in a wider sense, and see how general environmental factors can shape our health and lifespan.

How the Environment Affects Lifespan and Population Density

Here is a good place to discuss the r-K selection model of population dynamics. Living in a world where the environment is either risky or safe has profound implications on our reproductive strategies and our survival as individuals. A challenging environment is not necessarily a dangerous environment but it may help us survive if the stimuli remain within the hormetic (*i.e.* not too high and not too low) range. I will discuss some general aspects of the r-K selection model which are applicable to all species, but the intention is to show that humans who may be able to manipulate their environment, may also be able to modify the mechanisms and parameters of the model, and therefore affect their survival.

This model basically states that unpredictable and risky environments result in an r-type selection, whereas stable and safe environments characterise K-type selection [69]. In the r-type model, there are organisms with shorter lifespans compensated by large litters and early reproduction. These are generally less intelligent, with high sex drives. In the K model, organisms have longer lifespans, weak sex drive and small litters, with increased intelligence and slow rate of maturation [70] (Fig. **3**). Taking this model to the extreme, one can envisage a human society with very long-lived individuals who are intelligent, childless and, (despite the predictable unpopularity of the scenario) with low sex drive and a reduction of sexual activities (unless used to express social identity, or for recreation).

Starting from a premise based on the r-K selection model of population dynamics, it is assumed that in stable and safe environments, population stays stable with long lifespans and poor fertility with long-term development, whereas in unstable and dangerous environments there is increased fertility and shorter lifespans. In an uncertain and dangerous environment, organisms created high numbers of offspring, knowing that most will die, but at least increasing the chances that some will survive. In a safe environment this technique is not necessary as it would involve spending a lot of resources to bring up offspring which will not die. Therefore, there is a trade-off between resource use and number of offspring.

Increasing control over the environment (such as our current and continual technological influence over our environment) is associated with K-type selection and long term health.

RI (RATE OF INCREASE)

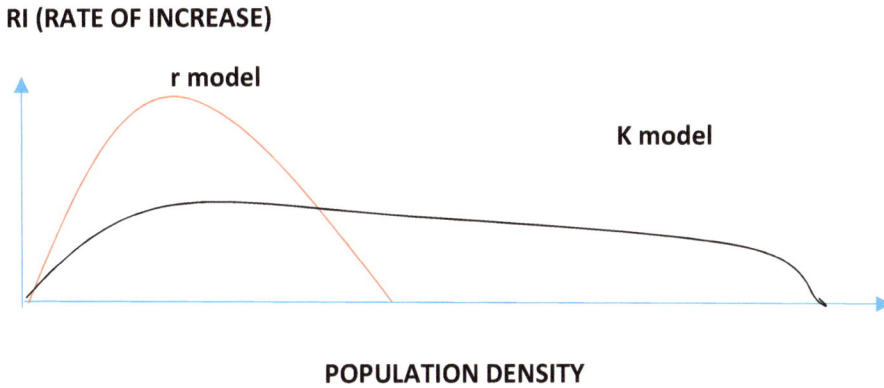

POPULATION DENSITY

Fig. (3). The r-K model. In the r-type model the population increases fast and decreases also fast. Instead, in the K-type model a modest but sustained population density is maintained with organisms experiencing longer lifespans.

DISCUSSION AND FUTURE PROSPECTS

I discussed the role of information, challenges and other environmental stimuli with regards to up-regulating certain important elements of our biological function. The mechanisms involved are various and sometimes not well-studied. Inevitably, there is a degree of informed speculation in some cases. However, the ultimate aim is not merely to study and analyse such mechanisms, but to see how these can be translated into practical curative methodologies that can be applied upon humanity at large, with the aim of reducing or eliminating age-related degeneration.

At this point it is imperative to distinguish between two terms which are causing considerable confusion in the field of healthy ageing: the distinction between a 'maximum lifespan' and an 'indefinite lifespan'. Suggestions that treatments and therapies currently exist, or may be devised in the near future, which may 'reverse ageing' or promote longevity, are abundant and conspicuously simplistic [71]. Some people, including scientists, indiscriminately use terms such as 'immortality' when what they really mean is healthy life extension until the age of

100-120 [72]. Others suggest biologically-sound rejuvenating therapies as a means for defying age-related degeneration and living longer [73], but without due consideration of the translational issues [74]. In my view, the concept of healthy human longevity has two distinct facets:

The first is the natural and existing developmental process, bound by Darwinian constrains which limit the maximum human lifespan to around 120 years. In this case, survival is ensured through germ-line immortalisation processes traded-off through a perpetual cycle of somatic birth, development, ageing and death, and an absolute maximum individual lifespan. Within this setting, human survival dynamics can be pliable and amenable to sometimes, simplistic, interventions. Examples include hormetins and nutritional factors, health supplements, antioxidants, cell repair strategies, telomerase interventions, stem cell therapies, genetic manipulation of relevant genes such as TOR, Foxo, Sir *etc.*, as well as other rejuvenation biotechnologies [75].

The second is encountered beyond the intrinsic barrier that limits human maximum lifespan. It is possible that, certain individuals will be subjected to natural evolutionary pressures that facilitate indefinite lifespan, in a transhumanist, post-Darwinian domain [76]. Here, 'indefinite lifespan' denotes a life free from constrains imposed by age-related degenerative loss of function, and without a pre-determined maximum age limit. In this situation, interventions based on physical therapeutic elements such as tablets or injections directed at the individual alone, will have a trivial impact upon longevity [77]. It would be necessary to devise strategies which are based on fundamental evolutionary principles [78] and encompassing wide ranging principles based on societal and planetary, if not universe-wide, approaches [79] within a framework of mutual and reciprocal co-operation. Within this domain, a highly sophisticated, complex approach will be needed in order to maintain a perpetual correction of somatic age-related damage, leading to a radically extended period with no significant age-related chronic degeneration [80].

As simple life extension interventions reach their peak, increasingly more people will live to attain the currently defined maximum lifespan. Beyond this limit, the importance of any physical rejuvenation and life extension therapies rapidly becomes less relevant, tending to triviality. At this stage, the relevance of radical life extension involving evolutionary-based techniques rapidly increases.

I suggest that the terms 'life extension' or 'longevity' are appropriate with regards to strategies aiming to prolong healthy life in order to reach the maximum lifespan limit. The terms 'radical life extension' or 'indefinite lifespan' should be used for strategies (whether idealistic or realistic) aiming to completely eliminate age-related degeneration and prolong lifespan beyond the current maximum limit. These two approaches are qualitatively different and distinct. For the avoidance of confusion, I emphasise once more that I define ageing as 'time-related loss of function'. This highlights the fact that the intention of the discussion is to elaborate on human ageing mechanisms and how these affect clinical degeneration.

Therefor our overall aim is:

a. First, to reduce the impact of age-related dysfunction and repair any associated clinical damage (through cures and therapies) and,

b. Second, to eventually achieve a state where age-related dysfunction becomes trivial (with a state of negligible senescence, where the mortality as a function of age becomes almost zero).

For the past several years I promoted my view that this second aim is quantitatively and qualitatively different from the first aim: Current and future biomedical and pharmacological technologies may help us achieve the first aim but not the second [81]. As mentioned above, my view is that these biomedical technologies will fail to be of any use with regards to our second aim [77].

Therefore we need to rely more on environmental, hormetic and evolutionarily-inspired concepts in order to achieve a more long-lasting and radical result. In addition, we need to consider fundamental science subjects such as the behaviour of complex adaptive systems, cybernetic principles, non-linearity and a host of others, some of which are discussed in this book. For instance, consider the case of a new emerging science: Molecular Pathological epidemiology (MPE). Research in this concept suggests that the influence of the environment on individual cells, organisms and populations is paramount [82]. MPE studies the relationship between the environment, cellular and extracellular processes and the evolution of pathology [83].

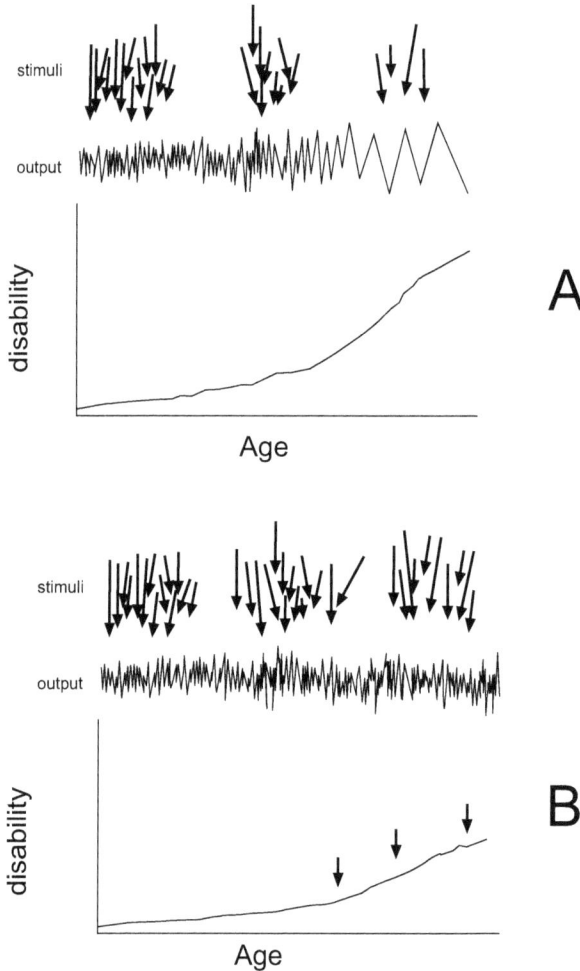

Fig. (4 (A and B)). The effects of multiple stimulation on disability and health.

A. During the early years of life an individual is exposed to a multitude of interacting stimuli which result in dynamically rich outputs, reflected in low levels of disability. From middle age and beyond, the amount of stimulation is reduced, the output signal gradually becomes less complex and more regular, and the level of disability increases.

B. Introducing and maintaining additional multiple interacting stimuli throughout life preserves the dynamical complexity of the output signal and, as a consequence, the level of disability is reduced. The individual lives a longer period with only minimal levels of disability, compared to the situation in Figure A. Eventually, disability does increase, but later on in life and not as significantly [From [86]].

There are elements of heterogeneity within cellular and organismic populations which cannot be treated as common to all individual members. The reference to the problems posed by interpersonal heterogeneity in pathology resonates with the concept of inter-individual variation encountered in hormesis (chapter 2), which shows that there are shared frontiers between pathology and hormesis [84]. Any disease evolution (including ageing, taken to mean 'time-related dysfunction') is a heterogeneous process, with unique patterns of disease in individuals. The outcome of each treatment is variable, as it depends on many parameters which are difficult or impossible to predict. Unique profiles of epigenomics, metabolomics and the macro and micro environments result in disease and ageing, but that this does not mean that deconstructing these profiles will necessarily give us an answer as to the treatment. Therefore it may be necessary to place more emphasis on personalised therapeutics, but any treatment methodologies which ignore this (and aim a general therapy for all), are likely to be unsuccessful [85].

CONCLUSION

A persistent underlying theme throughout this discussion is based on concepts drawn not only from the discipline of hormesis, but also from evolutionary theory. Modern evolution is underpinned by the notion of 'co-operation' (not competition) leading to increased fitness within a progressively more technologically advanced niche. Those who are able to adapt to this, will succeed in surviving, and will surviving longer compared to those who choose competition as a strategy for survival. Adapting well to novel environmental stimulation is paramount. Those who fail to respond to the challenges found in an increasingly technocratic society are likely to experience a decrease in fitness, and consequently their survival rates will decline (Fig. **4**). The answer to eliminating ageing must lie in the study of more sophisticated concepts, based on our understanding of how exactly the environment affects our biological processes, what our environment actually is, and how we can intentionally react to the continual exposure to challenges originating from a progressively technological environment.

REFERENCES

[1]   Garthe A, Roeder I, Kempermann G. Mice in an enriched environment learn more flexibly because of

adult hippocampal neurogenesis. Hippocampus 2015; ••• [Epub ahead of print].
[http://dx.doi.org/10.1002/hipo.22520] [PMID: 26311488]

[2]   Birch AM, McGarry NB, Kelly AM. Short-term environmental enrichment, in the absence of exercise, improves memory, and increases NGF concentration, early neuronal survival, and synaptogenesis in the dentate gyrus in a time-dependent manner. Hippocampus 2013; 23(6): 437-50.
[http://dx.doi.org/10.1002/hipo.22103] [PMID: 23460346]

[3]   Bayne K, Würbel H. The impact of environmental enrichment on the outcome variability and scientific validity of laboratory animal studies. Rev - Off Int Epizoot 2014; 33(1): 273-80.
[http://dx.doi.org/10.20506/rst.33.1.2282] [PMID: 25000800]

[4]   Mora-Gallegos A, Rojas-Carvajal M, Salas S, Saborío-Arce A, Fornaguera-Trías J, Brenes JC. Age-dependent effects of environmental enrichment on spatial memory and neurochemistry. Neurobiol Learn Mem 2015; 118: 96-104.
[http://dx.doi.org/10.1016/j.nlm.2014.11.012] [PMID: 25434818]

[5]   Mora F, Segovia G, del Arco A. Aging, plasticity and environmental enrichment: structural changes and neurotransmitter dynamics in several areas of the brain. Brain Res Brain Res Rev 2007; 55(1): 78-88.
[http://dx.doi.org/10.1016/j.brainresrev.2007.03.011] [PMID: 17561265]

[6]   Kelly ÁM. Non-pharmacological Approaches to Cognitive Enhancement. Handbook Exp Pharmacol 2015; 228: 417-39.
[http://dx.doi.org/10.1007/978-3-319-16522-6_14] [PMID: 25977091]

[7]   Pang TY, Hannan AJ. Enhancement of cognitive function in models of brain disease through environmental enrichment and physical activity. Neuropharmacology 2013; 64: 515-28.
[http://dx.doi.org/10.1016/j.neuropharm.2012.06.029] [PMID: 22766390]

[8]   Grinde B, Patil GG. Biophilia: does visual contact with nature impact on health and well-being? Int J Environ Res Public Health 2009; 6(9): 2332-43.
[http://dx.doi.org/10.3390/ijerph6092332] [PMID: 19826546]

[9]   Capaldi CA, Dopko RL, Zelenski JM. The relationship between nature connectedness and happiness: a meta-analysis. Front Psychol 2014; 5: 976.
[http://dx.doi.org/10.3389/fpsyg.2014.00976] [PMID: 25249992]

[10]   Grady DL, Thanos PK, Corrada MM, *et al.* DRD4 genotype predicts longevity in mouse and human. J Neurosci 2013; 33(1): 286-91.
[http://dx.doi.org/10.1523/JNEUROSCI.3515-12.2013] [PMID: 23283341]

[11]   van Strien T, Levitan RD, Engels RC, Homberg JR. Season of birth, the dopamine D4 receptor gene and emotional eating in males and females. Evidence of a genetic plasticity factor? Appetite 2015; 90: 51-7.
[http://dx.doi.org/10.1016/j.appet.2015.02.024] [PMID: 25736807]

[12]   van der Wal AJ, Schade HM, Krabbendam L, van Vugt M. Do natural landscapes reduce future discounting in humans? Proc Biol Sci 2013; 280(1773)

[13]   Pati D, Freier P, O'Boyle M, Amor A, Valipoor S. The Impact of Simulated Nature on Patient Outcomes: A Study of Photographic Sky Compositions. HERD 2015; 1937586715595505. [Epub

ahead of print].
[PMID: 26199272]

[14]   Pergams OR, Zaradic PA. Is love of nature in the US becoming love of electronic media? 16-year downtrend in national park visits explained by watching movies, playing video games, internet use, and oil prices. J Environ Manage 2006; 80(4): 387-93.
       [http://dx.doi.org/10.1016/j.jenvman.2006.02.001] [PMID: 16580127]

[15]   Kyriazis M. Technological integration and hyperconnectivity: Tools for promoting extreme human lifespans. Complexity 2015; 20(6): 15-24.
       [http://dx.doi.org/10.1002/cplx.21626]

[16]   Gong D, He H, Liu D, *et al.* Enhanced functional connectivity and increased gray matter volume of insula related to action video game playing. Sci Rep 2015; 5: 9763.
       [http://dx.doi.org/10.1038/srep09763] [PMID: 25880157]

[17]   Oei AC, Patterson MD. Enhancing cognition with video games: a multiple game training study. PLoS One 2013; 8(3): e58546.
       [http://dx.doi.org/10.1371/journal.pone.0058546] [PMID: 23516504]

[18]   Liang T-H. Association between Use of internet services and quality of life in Taiwan. J Data Sci 2011; 9: 83-92.

[19]   Jones RB, Ashurst EJ, Atkey J, Duffy B. Older people going online: its value and before-after evaluation of volunteer support. J Med Internet Res 2015; 17(5): e122.
       [http://dx.doi.org/10.2196/jmir.3943] [PMID: 25986724]

[20]   Kyriazis M. Evolutionary replicators in the Global Brain. Available from: http://biologicalimmortality.blogspot.co.uk/2011_03_01_archive.html 2011.

[21]   Kyriazis M. Systems neuroscience in focus: from the human brain to the global brain? Frontiers Syst Neurosci 2015; 9(7) eCollection
       [http://dx.doi.org/10.3389/fnsys.2015.00007]

[22]   Graham A. The Digital Physical Blur in Our Digital Business Era. Available from: http://info.exsquared.com/ex-squared-blog/the-digital-physical-blur-in-our-digital-business-era 2015.

[23]   Lis E, Chiniara C, Wood MA, Biskin R, Montoro R. Psychiatrists' Perceptions of World of Warcraft and Other MMORPGs. Psychiatr Q 2015. [Epub ahead of print].
       [http://dx.doi.org/10.1007/s11126-015-9358-2] [PMID: 26275869]

[24]   Seo JH, Kim H, Park ES, *et al.* Environmental enrichment synergistically improves functional recovery by transplanted adipose stem cells in chronic hypoxic-ischemic brain injury. Cell Transplant 2013; 22(9): 1553-68.
       [http://dx.doi.org/10.3727/096368912X662390] [PMID: 23394350]

[25]   Seo JH, Yu JH, Suh H, Kim MS, Cho SR. Fibroblast growth factor-2 induced by enriched environment enhances angiogenesis and motor function in chronic hypoxic-ischemic brain injury. PLoS One 2013; 8(9): e74405.
       [http://dx.doi.org/10.1371/journal.pone.0074405] [PMID: 24098645]

[26]   Steventon JJ, Harrison DJ, Trueman RC, Rosser AE, Jones DK, Brooks SP. *In Vivo* MRI evidence that neuropathology is attenuated by cognitive enrichment in the Yac128 huntington's disease mouse

model. J Huntingtons Dis 2015; 4(2): 149-60.
[http://dx.doi.org/10.3233/JHD-150147] [PMID: 26397896]

[27]   Heylighen F. Cybernetic principles of aging and rejuvenation: the buffering- challenging strategy for life extension. Curr Aging Sci 2014; 7(1): 60-75.
[http://dx.doi.org/10.2174/1874609807666140521095925] [PMID: 24852018]

[28]   Fong P, Forster P. The Social Benefits of Computer Games Proceedings of the 44th Annual APS Conference.   Available   from: https://www.academia.edu/705294/The_Social_Benefits_of_Computer_Games 2009; 62-5.

[29]   Pack Kaelbling L, Littman M, Moore A. Reinforcement Learning: A Survey. J Artif Intell Res 1996; 4: 237-85.

[30]   Teng CY, Adamic L. Longevity in Second Life. Proceedings of ICWSM.

[31]   Brandtzaeg PB, Heim J. User loyalty and online communities: why members of online communities are not faithful. INTETAIN 2008; pp. 1-10.

[32]   Ramírez-Rodríguez G, Ocaña-Fernández MA, Vega-Rivera NM, *et al.* Environmental enrichment induces neuroplastic changes in middle age female Balb/c mice and increases the hippocampal levels of BDNF, p-Akt and p-MAPK1/2. Neuroscience 2014; 260: 158-70.
[http://dx.doi.org/10.1016/j.neuroscience.2013.12.026] [PMID: 24361917]

[33]   Kyriazis   M.   The   law   of   requisite   usefulness.   Available   from: http://scienceblog.com/76141/law-requisite-usefulness/ 2014.

[34]   Kyriazis M. Practical applications of chaos theory to the modulation of human ageing: nature prefers chaos to regularity. Biogerontology 2003; 4(2): 75-90.
[http://dx.doi.org/10.1023/A:1023306419861] [PMID: 12766532]

[35]   Kyriazis M. Nonlinear stimulation and hormesis in human aging: practical examples and action mechanisms. Rejuvenation Res 2010; 13(4): 445-52.
[http://dx.doi.org/10.1089/rej.2009.0996] [PMID: 20662589]

[36]   Takano T, Nakamura K, Watanabe M. Urban residential environments and senior citizens' longevity in megacity areas: the importance of walkable green spaces. J Epidemiol Community Health 2002; 56(12): 913-8.
[http://dx.doi.org/10.1136/jech.56.12.913] [PMID: 12461111]

[37]   Bittman B, Berk L, Shannon M, *et al.* Recreational music-making modulates the human stress response: a preliminary individualized gene expression strategy. Med Sci Monit 2005; 11(2): BR31-40.
[PMID: 15668624]

[38]   Kanduri C, Raijas P, Ahvenainen M, *et al.* The effect of listening to music on human transcriptome. PeerJ 2015; 3: e830.
[http://dx.doi.org/10.7717/peerj.830] [PMID: 25789207]

[39]   Thornley J, Hirjee H, Vasudev A. Music therapy in patients with dementia and behavioral disturbance on an inpatient psychiatry unit: results from a pilot randomized controlled study. Int Psychogeriatr 2015; 1-3. [Epub ahead of print].
[PMID: 26572722]

[40]   Pedersen E. City dweller responses to multiple stressors intruding into their homes: noise, light, odour, and vibration. Int J Environ Res Public Health 2015; 12(3): 3246-63.
[http://dx.doi.org/10.3390/ijerph120303246] [PMID: 25794188]

[41]   Kang JI, Song DH, Namkoong K, Kim SJ. Interaction effects between COMT and BDNF polymorphisms on boredom susceptibility of sensation seeking traits. Psychiatry Res 2010; 178(1): 132-6.
[http://dx.doi.org/10.1016/j.psychres.2010.04.001] [PMID: 20434221]

[42]   Wood W, Womack J, Hooper B. Dying of boredom: an exploratory case study of time use, apparent affect, and routine activity situations on two Alzheimer's special care units. Am J Occup Ther 2009; 63(3): 337-50.
[http://dx.doi.org/10.5014/ajot.63.3.337] [PMID: 19522143]

[43]   Buchholz L. Exploring the promise of mindfulness as medicine. JAMA 2015; 314(13): 1327-9.
[http://dx.doi.org/10.1001/jama.2015.7023] [PMID: 26441167]

[44]   Lutz J, Brühl AB, Doerig N, *et al.* Altered processing of self-related emotional stimuli in mindfulness meditators. Neuroimage 2015; 124(Pt A): 958-67.

[45]   Taylor VA, Daneault V, Grant J, *et al.* Impact of meditation training on the default mode network during a restful state. Soc Cogn Affect Neurosci 2013; 8(1): 4-14.
[http://dx.doi.org/10.1093/scan/nsr087] [PMID: 22446298]

[46]   Johnson JR, Emmons HC, Rivard RL, Griffin KH, Dusek JA. Resilience Training: A Pilot Study of a Mindfulness-Based Program With Depressed Healthcare Professionals Explore (NY) 2015; S1550-8307(15)

[47]   Fraga MF, Ballestar E, Paz MF, *et al.* Epigenetic differences arise during the lifetime of monozygotic twins. Proc Natl Acad Sci USA 2005; 102(30): 10604-9.
[http://dx.doi.org/10.1073/pnas.0500398102] [PMID: 16009939]

[48]   Dolinoy DC, Weinhouse C, Jones TR, Rozek LS, Jirtle RL. Variable histone modifications at the Avy metastable epiallele. Epigenetics: official journal of the DNA Methylation Society 2010; 5(7): 637-44.

[49]   Lansade L, Valenchon M, Foury A, *et al.* Behavioral and transcriptomic fingerprints of an enriched environment in horses (*Equus caballus*). PLoS One 2014; 9(12): e114384.
[http://dx.doi.org/10.1371/journal.pone.0114384] [PMID: 25494179]

[50]   Sheikhzadeh F, Etemad A, Khoshghadam S, Asl NA, Zare P. Hippocampal BDNF content in response to short- and long-term exercise. Neurol Sci 2015; 36(7): 1163-6.
[http://dx.doi.org/10.1007/s10072-015-2208-z] [PMID: 25860428]

[51]   Hochachka PW. Solving the common problem: matching ATP synthesis to ATP demand during exercise. Adv Vet Sci Comp Med 1994; 38A: 41-56.
[PMID: 7801835]

[52]   Morse SJ, Butler AA, Davis RL, Soller IJ, Lubin FD. Environmental enrichment reverses histone methylation changes in the aged hippocampus and restores age-related memory deficits. Biology (Basel) 2015; 4(2): 298-313.
[http://dx.doi.org/10.3390/biology4020298] [PMID: 25836028]

[53] Frick KM. Epigenetics, oestradiol and hippocampal memory consolidation. J Neuroendocrinol 2013; 25(11): 1151-62.
[http://dx.doi.org/10.1111/jne.12106] [PMID: 24028406]

[54] Kempermann G, Gast D, Gage FH. Neuroplasticity in old age: sustained fivefold induction of hippocampal neurogenesis by long-term environmental enrichment. Ann Neurol 2002; 52(2): 135-43.
[http://dx.doi.org/10.1002/ana.10262] [PMID: 12210782]

[55] Anonymous . Gut microbiota. Available from: http://www.nature.com/nrgastro/focus/gutmicrobiota/index.html 2015.

[56] Amato KR, Leigh SR, Kent A, *et al.* The gut microbiota appears to compensate for seasonal diet variation in the wild black howler monkey (Alouatta pigra). Microb Ecol 2015; 69(2): 434-43.
[http://dx.doi.org/10.1007/s00248-014-0554-7] [PMID: 25524570]

[57] Zanotti I, Turroni F, Piemontese A, *et al.* Evidence for cholesterol-lowering activity by *Bifidobacterium bifidum* PRL2010 through gut microbiota modulation. Appl Microbiol Biotechnol 2015; 99(16): 6813-29.
[http://dx.doi.org/10.1007/s00253-015-6564-7] [PMID: 25863679]

[58] Kim TH, Yang J, Darling PB, O'Connor DL. A large pool of available folate exists in the large intestine of human infants and piglets. J Nutr 2004; 134(6): 1389-94.
[PMID: 15173401]

[59] O'Keefe SJ. Nutrition and colonic health: the critical role of the microbiota. Curr Opin Gastroenterol 2008; 24(1): 51-8.
[http://dx.doi.org/10.1097/MOG.0b013e3282f323f3] [PMID: 18043233]

[60] Olson KR. The therapeutic potential of hydrogen sulfide: separating hype from hope. Am J Physiol Regul Integr Comp Physiol 2011; 301(2): R297-312.
[http://dx.doi.org/10.1152/ajpregu.00045.2011] [PMID: 21543637]

[61] Kashfi K, Olson KR. Biology and therapeutic potential of hydrogen sulfide and hydrogen sulfide-releasing chimeras. Biochem Pharmacol 2013; 85(5): 689-703.
[http://dx.doi.org/10.1016/j.bcp.2012.10.019] [PMID: 23103569]

[62] Ridaura V, Belkaid Y. Gut microbiota: the link to your second brain. Cell 2015; 161(2): 193-4.
[http://dx.doi.org/10.1016/j.cell.2015.03.033] [PMID: 25860600]

[63] Yano JM, Yu K, Donaldson GP, *et al.* Indigenous bacteria from the gut microbiota regulate host serotonin biosynthesis. Cell 2015; 161(2): 264-76.
[http://dx.doi.org/10.1016/j.cell.2015.02.047] [PMID: 25860609]

[64] Aidy SE, Dinan TG, Cryan JF. The Conductor in the Orchestra of Immune-Neuroendocrine Communication. Clin Ther 2015; S0149-2918(15)

[65] O'Sullivan E, Barrett E, Grenham S, *et al.* BDNF expression in the hippocampus of maternally separated rats: does *Bifidobacterium breve* 6330 alter BDNF levels? Benef Microbes 2011; 2(3): 199-207.
[http://dx.doi.org/10.3920/BM2011.0015] [PMID: 21986359]

[66] Hutton CP, Déry N, Rosa E, *et al.* Synergistic effects of diet and exercise on hippocampal function in chronically stressed mice. Neuroscience 2015; 308: 180-93.

[http://dx.doi.org/10.1016/j.neuroscience.2015.09.005] [PMID: 26358368]

[67]   Rahe J, Becker J, Fink GR, *et al.* Cognitive training with and without additional physical activity in healthy older adults: cognitive effects, neurobiological mechanisms, and prediction of training success. Front Aging Neurosci 2015; 7: 187.
[http://dx.doi.org/10.3389/fnagi.2015.00187] [PMID: 26528177]

[68]   Dinan TG, Stilling RM, Stanton C, Cryan JF. Collective unconscious: how gut microbes shape human behavior. J Psychiatr Res 2015; 63: 1-9.
[http://dx.doi.org/10.1016/j.jpsychires.2015.02.021] [PMID: 25772005]

[69]   Pianka ER. On r and K selection. Am Nat 1970; 104(940): 592-7.
[http://dx.doi.org/10.1086/282697]

[70]   Parry GD. The Meanings of r- and K-selection. Oecologia 1981; 48(2): 260-4.
[http://dx.doi.org/10.1007/BF00347974]

[71]   Kyriazis M. No Silver Bullets for Aging - Alchemy *vs* Networks. H+ Magazine. Available from: http://hplusmagazine.com/2014/12/10/no-silver-bullets-aging-alchemy-vs-networks/ 2015.

[72]   Kiefer B. [Immortality, death and other minor phenomena]. Rev Med Suisse 2015; 11(477): 1264.
[PMID: 26211294]

[73]   de Grey AD. Longevity Sticker Shock: The One Remaining Obstacle to Widespread Credentialed Support for Rejuvenation Biotechnology. Rejuvenation Res 2015; 18(3): 201-2.
[http://dx.doi.org/10.1089/rej.2015.1732] [PMID: 26053409]

[74]   Kyriazis M. Translating laboratory anti-aging biotechnology into applied clinical practice: Problems and obstacles. World J Transl Med 2015; 4(2): 51-4.
[http://dx.doi.org/10.5528/wjtm.v4.i2.51]

[75]   de Magalhães JP. The scientific quest for lasting youth: prospects for curing aging. Rejuvenation Res 2014; 17(5): 458-67.
[http://dx.doi.org/10.1089/rej.2014.1580] [PMID: 25132068]

[76]   Kyriazis M. Reversal of informational entropy and the acquisition of germ-like immortality by somatic cells. Curr Aging Sci 2014; 7(1): 9-16.
[http://dx.doi.org/10.2174/1874609807666140521101102] [PMID: 24852017]

[77]   Kyriazis M. The impracticality of biomedical rejuvenation therapies: translational and pharmacological barriers. Rejuvenation Res 2014; 17(4): 390-6.
[http://dx.doi.org/10.1089/rej.2014.1588] [PMID: 25072550]

[78]   Heylighen F. Cybernetic principles of aging and rejuvenation: the buffering- challenging strategy for life extension. Curr Aging Sci 2014; 7(1): 60-75.
[http://dx.doi.org/10.2174/1874609807666140521095925] [PMID: 24852018]

[79]   Vidal C. Cosmological immortality: how to eliminate aging on a universal scale. Curr Aging Sci 2014; 7(1): 3-8.
[http://dx.doi.org/10.2174/1874609807666140521111107] [PMID: 24852011]

[80]   Kyriazis M. Editorial: Novel approaches to an old problem: insights, theory and practice for eliminating aging. Curr Aging Sci 2014; 7(1): 1-2.
[http://dx.doi.org/10.2174/1874609807011407031030943] [PMID: 25056407]

[81]   Kyriazis M, Apostolides A. The Fallacy of the Longevity Elixir: Negligible Senescence May be Achieved, but Not by Using Something Physical. Curr Aging Sci 2015; 8(3): 227-34.
[http://dx.doi.org/10.2174/1874609808666150702095803] [PMID: 26135528]

[82]   Ogino S, Lochhead P, Chan AT, *et al.* Molecular pathological epidemiology of epigenetics: emerging integrative science to analyze environment, host, and disease. Mod Pathol 2013; 26(4): 465-84.
[http://dx.doi.org/10.1038/modpathol.2012.214] [PMID: 23307060]

[83]   Ogino S, Campbell PT, Nishihara R, *et al.* Proceedings of the second international molecular pathological epidemiology (MPE) meeting. Cancer Causes Control 2015; 26(7): 959-72.
[http://dx.doi.org/10.1007/s10552-015-0596-2] [PMID: 25956270]

[84]   Nishihara R, VanderWeele TJ, Shibuya K, *et al.* Molecular pathological epidemiology gives clues to paradoxical findings. Eur J Epidemiol 2015; 30(10): 1129-35.
[http://dx.doi.org/10.1007/s10654-015-0088-4] [PMID: 26445996]

[85]   Nishi A, Kawachi I, Koenen KC, Wu K, Nishihara R, Ogino S. Lifecourse epidemiology and molecular pathological epidemiology. Am J Prev Med 2015; 48(1): 116-9.
[http://dx.doi.org/10.1016/j.amepre.2014.09.031] [PMID: 25528613]

[86]   Kyriazis M. Practical applications of chaos theory to the modulation of human ageing: nature prefers chaos to regularity. Biogerontology 2003; 4(2): 75-90.
[http://dx.doi.org/10.1023/A:1023306419861] [PMID: 12766532]

CHAPTER 4

Epigenetic Regulation and Adaptation to Stimuli

Abstract: Changes and variations to physiological traits which are caused by environmental factors are studied by the science of epigenetics. This sub-section of genetics describes alterations in transcription which may result in different phenotypes depending on the influence of the environment. The study of epigenetic mechanisms is very relevant in ageing and, in particular, in situations involving external challenges and exposure to novel information. In this chapter, I will discuss some elements related to epigenetic regulation as applied to situations where humans are exposed to 'positive challenges' which aim to up-regulate somatic repair mechanisms. The role of epigenetic factors, including non-coding RNAs (such as microRNAs) will also be discussed in the context of an information-rich environment. In addition, I will explore certain principles (such as the Condition-Action rule) which are relevant to the operation of the adaptation mechanisms, and some mechanisms which govern the process of information assimilation by the cell. The underlying theme, which will also be explored in the following chapters, is the consideration of mechanisms that may result in reallocation of repair resources from the germ line back to the soma. Overall, the aim is to highlight factors, processes and principles which depend on the environment and may be involved in human health improvement.

Keywords: Adaptation, Condition-Action rule, Environmental challenges, Epigenetic landscape, Epigenetic regulation External stimulation, Germ line repair, Information, MicroRNAs, Soma-germ line conflict.

INTRODUCTION

Adaptive phenotypic plasticity is an important factor in human evolution. There is considerable evidence suggesting that the environment influences the plasticity of the phenotype [1] and that there exist precise, if hitherto not very well studied, mechanisms, which respond to stimulating changes in the environment, in an attempt to adapt to the stimulus [2]. This adaptation to a stimulus is very relevant

Marios Kyriazis

in the study of age-related pathology and in the evolution of ageing in general. It is also relevant in the context of hormesis and stress response processes. Increasingly, research shows that the environment is crucial in determining prolonged healthy longevity [3]. In human terms, the 'environment' is an abstract notion of an amalgam of physical and virtual surroundings, interactions with modern society, and techno-cultural elements. It is, in other words, a highly 'information-rich' milieu. It is worth noting at this point that I define information as 'a meaningful set of data or patterns which influence the formation or transformation of other data or patterns, in order to reduce uncertainty and help achieve a goal' [4] (Fig. **1**).

Fig. (1). The impact of information upon survival. Information (plus organisation) [5] increases complexity and this increases functionality [6]. This improves fitness and thus survival [7]. Any increase in internal fitness requires the formation of new links and the strengthening of the interconnectedness between its nodes *i.e.* increased complexity, thus increased fitness and increased survival. The definition of **fitness** is 'good function within a specified environment'.

Ashby's law of Requisite Variety [8] suggests that less predictable environments (where we don't know what will happen next, *i.e.* those environments that are more challenging) require more complexity of function in order to survive in that specific environment. As our technological environment is now increasing in complexity and becomes more 'information-rich', we too need to adapt to these changes in order to function efficiently and survive. Our technology and evolving culture are driving us to continually adapt and change, in a perpetual attempt to become better able to deal with new challenges. However, there has to be a balance both with respect to the amount of stimulation and with regards to an optimum level of unpredictability. If this balance is disturbed for significant periods of time, then overexposure to information will result in loss of function. This is a 'dose-response' phenomenon, *i.e.* a hormetic effect. The concept of 'information fatigue' becomes increasingly more relevant in a society which is inundated in information.

In addition, as our environment is constantly changing, (and recently, changing at an ever-accelerating manner [9]) we need to adapt to these changes quite fast. This must depend on intentional evolutionary actions executed through technology rather than the much slower Darwinian natural selection processes [10]. There is simply not enough time for us to continue relying on natural selection, when new disruptive and global technologies appear so quickly. Epigenetic factors mediate between the genome and the environment, and play an important role in this respect, allowing for rapid changes in protein expression which may be applied almost immediately when needed. Of course, there are other factors which may initiate and achieve rapid biological and physiological adaptations, but the epigenetic ones are the most relevant.

Epigenetic Regulation in Ageing

There is an increasing body of research which highlights the role of epigenetic changes in age-related degeneration [11]. Epigenetic changes such as chromatin remodelling and DNA methylation can modify the phenotype and play a modulating role in senescence and age-disease. One of the reasons for this, is based on the fact that epigenetic processes can regulate mechanisms necessary for development [12]. Ageing has been interpreted by several authors as a continual and disproportionate operation of basic developmental processes, which eventually lead to time-related malfunction [13, 14]. Therefore, a corrective regulation of this defective process could lead to the control of at least some ageing mechanisms. In addition, it is known that there could be 'molecular brakes' which restrict the plasticity of adult cells in certain types of neurons [15]. These and other epigenetic mechanisms may lead to stabilisation and delayed expression of any brake factors and thus account for seamless and rapid plasticity of the cell. The fact that several of these factors are found in the flexible and adaptable extracellular space makes it more likely that epigenetic control would be more effective compared to 'hard-wired' genetic intracellular mechanisms.

Epigenetic Changes and Hormesis: The Epigenetic Landscape

Referring to the complexity of gene regulatory networks and to the effect of the environment, Huang [16] remarked:

The Neo-Darwinian concept of natural selection is plausible when one assumes a straightforward causation of phenotype by genotype. However, such simple 1:1 mapping must now give place to the modern concepts of gene regulatory networks and gene expression noise. Both can, in the absence of genetic mutations, jointly generate a diversity of inheritable randomly occupied phenotypic states that could also serve as a substrate for natural selection. This form of epigenetic dynamics challenges Neo-Darwinism. It needs to incorporate the non-linear, stochastic dynamics of gene networks. A first step is to consider the mathematical correspondence between gene regulatory networks and Waddington's metaphoric 'epigenetic landscape', which actually represents the quasi-potential function of global network dynamics. It explains the coexistence of multiple stable phenotypes within one genotype. The landscape's topography with its attractors is shaped by evolution through mutational re-wiring of regulatory interactions - offering a link between genetic mutation and sudden, broad evolutionary changes.

This highlights the huge importance of taking into account the environment where an organism is in, and also the nature of interactions with this environment. Our actions and general fitness are not defined by genes alone but by a complex, cross-talking set of interactions where the epigenome, the gene expression mechanisms and other epigenetic factors play an important part, forming a dynamic, ever-changing and adaptable 'epigenetic landscape'. It is tempting to speculate that the environment will prove the definitive link between our current genetic inheritance (*i.e.* genes which express germ line-but not somatic repair factors) and our future evolutionary development, a fast and wide-ranging change in our biology (whereby the reverse becomes true: expression of somatic-but not germ line repair factors), as discussed in the following chapters.

The fact that the environment affects epigenetic regulatory mechanisms has been shown in many other studies [17, 18]. Several environmental factors are closely linked to the epigenetic regulation of gene expression. This epigenetic regulation may lead to adaptive changes, which increase the likelihood of successful survival within a particular niche. The epigenetic changes can be predictable, can occur both during early and late developmental stages, and can result in improvement of the functional ability of the organism [19].

It has been shown that environmental stress induces epigenetic changes and results in an adaptive response to the specific stressful event in question [20]. The question arises: "what is stress?" In the context of this book, 'stress' is taken to mean exposure to a stimulus which disrupts homeostasis and necessitates a reaction in order for the organism to adapt to this stimulus. I need to highlight once more that excessive, prolonged or severe stress does not come under the intended definition of stress for health purposes. As described in detail in Chapters 1 and 2, 'stress' (or more colloquially 'positive stress') must be of hormetic origin *i.e.* a dose-response phenomenon (with low-dose stimulation and high-dose inhibition) [21 - 25]. Vaiserman [26] considers that a mild stress-induced hormetic response involves mechanisms similar to those mechanisms that underlie developmental epigenetic adaptations, with resulting life-span prolonging properties. Examples of positive hormetic stress include those discussed in previous chapters, such as calorie restriction, physical exercise, and certain chemical compounds including free radicals and mild toxins. In addition, there could be other types of hormetic stimulation such as cultural, societal and psychological factors, all of which interact and integrate at multiple levels. It can be argued that hormetic stress, as the type encountered in our everyday highly technological, intensely information-sharing environment, is operating along conventional mechanisms that activate stress response pathways, and may cause direct epigenetic changes, which facilitate our adaptation to this environment. But, by scrutinising these epigenetic changes more closely, one may find that there are similarities and shared frontiers with some of the mechanisms that repair age-related damage. This may be due to rapid up-regulation of genes responsible for adaptive response, which can have an effect on healthy life extension [27]. Although excessive (prolonged and of high magnitude) stress is a well-known determinant for disease, I have already discussed in chapter 1 some ways by which we can decide if the degree of stress is likely to be beneficial or detrimental.

Even though epigenetic adaptations to a changing environment can occur during early stages of development, it is also possible to experience epigenetic plasticity in later life stages [28]. This is a promising fact, which supports the view that hormetic stressors can up-regulate healthy longevity mechanisms even if applied later in life. Although the search for an ideal type of mild positive (hormetic) stress still continues, it makes sense to consider stimuli which are directly relevant to our current technological and information-rich environment; stresses that must lead to adaptation. For instance, exposure to a persistent information-sharing environment, such as **meaningful** online hyper-connectedness, which acts as a kind of hormetic stimulus would lead to an adaptive response which may up-regulate several health parameters [29] as discussed below. Here, it must be emphasized that the exposure refers to meaningful stimulation containing information that requires action, and not trivial or meaningless information. Trivial information such as that we encounter in our everyday routine does not qualify as hormetic. In addition, one must be mindful of 'information fatigue' when information overload may overwhelm our cognitive system. Information that requires action operates as a challenge which rearranges the different elements within the epigenetic landscape resulting in adaptation, improvement and readiness for further stimulation, *i.e.* successful survival within a technological society.

The Case of Ageing: Somatic Cell Repair and Epigenetic Mechanisms

How can an information-rich environment influence biological parameters and lead to a reduced rate of age-related dysfunction? Here we need to consider the process of ageing in the context of both epigenetics and evolution. It has been suggested that although efficient repair mechanisms are present in somatic cells, these are traded-off in germ-line cells against optimal and successful survival (of the species) [30]. Heininger [31] quotes:

> *Evolution through natural selection can be described as driven by a perpetual conflict of individuals competing for limited resources.* ***Somatic death is a germ cell-triggered event*** *(emphasis mine) and has been established as evolutionary-fixed default state following asymmetric reproduction in a world of finite resources. Aging, on the*

other hand, is the stress resistance-dependent phenotype of the somatic resilience that counteracts the germ cell-inflicted death pathway. Thus, aging is a survival response and, in contrast to current beliefs, is antagonistically linked to death that is not imposed by group selection but enforced upon the soma by the selfish genes of the "enemy within" (i.e. the germ-line cells - comment and emphasis mine*). Environmental conditions shape the trade-off solutions as compromise between the conflicting germ–soma interests. Mechanistically, the neuroendocrine system, particularly those components that control energy balance, reproduction and stress responses, orchestrate these events.*

Although these comments have been made over a decade ago, research is now providing empirical and experimental support to this general notion. Examples of this research will be discussed below. Therefore, the concept of ageing can be viewed, at least partly, as a conflict between germ-line cells and somatic cells, whereby immortalisation factors are preferentially diverted towards the germ-line in order to assure the survival of the species, instead of being used by the somatic cells [32]. This results in a continual effective repair of the germ cells, which exhibit robust anti-senescence traits, at the expense of the somatic cells which eventually become overwhelmed by chronic damage.

Here it is worth mentioning that although the germ line may indeed exhibit certain characteristics of senescence over time, the overall result is that deleterious mutations and other germ line damage are eliminated or made safe by a steady and sufficient flow of repair resources as mentioned above. It is a kind of maintenance gap (trade-off) between the germ line and the soma. The resulting maintenance gap is a matter of both resources required and resources allocated, that is to say that if resources are constantly required, then these will be allocated in order to assure long term survival, despite the short-term evolutionary cost [33]. This general concept of 'consequence-capture' describes how agents capture sufficient of the benefits of their actions on other systems. If any given action creates benefits for the rest of the system, benefits which exceed the energy costs of those actions, then the agent will capture sufficient of those benefits to outweigh its own costs, and thus it will go ahead and take those actions. In other words, if the benefits of ongoing somatic repair are applicable to the entire species, then the actions (the somatic repair actions) leading to these benefits will be executed. I will return to this topic in later chapters.

The main existing evolutionary reason for the current soma-germ line conflict is the need to assure the survival of the species, which can currently be achieved much more energy-efficiently through the increased repair fidelity in germ line cells. I have suggested that it may be possible to re-balance (or reverse) the germ-soma conflict by using continual exposure to insightful information, a mechanism which depends on fast epigenetic modifications and is in accordance with basic evolutionary requirements [5].

Living in a highly technological society is a typical example of how environmental challenges (*i.e.* hyper-connection and a continual exposure to digital technology, for example) can have a direct impact upon age-regulating mechanisms *via* epigenetic instead of genetic changes. It is possible to examine further the role of epigenetic mechanisms in relation to the shifting of resources from germ line to somatic repairs. For instance, it is known that there exist non-autonomous contributions of somatic cells to germ line cells, leading to germ line immortalisation [34]. These contributions aid germ line survival, at the expense of somatic survival. Such strategies may also be present in somatic cells, but are significantly down-regulated [35]. Gracida and Eckmann [36] have uncovered an unprecedented and evolutionarily-conserved soma-to germ line communication pathway whereby somatic nuclear receptors (such as the *nhr-114* receptor) buffer against toxic dietary metabolites and actively protect germ line stem cells. This somatic nuclear receptor acts as a detoxifier and shields germ line stem cells from negative environmental challenges [37].

The aim here is to examine how this general rejuvenation process can be driven to operate in somatic cells [38], and how the trade-offs between somatic and germ line repairs can be reversed. It is important to highlight that certain mechanisms of germ line rejuvenation could be dependent upon epigenetic modifications and factors that regulate transcription [39]. Thus, it is also plausible that epigenetic co-ordination may have certain rejuvenating effects upon somatic cells although, clearly, more research is needed in this area. It is also worth remembering that not all epigenetic regulation may lead to positive results. For example, increasing methylation may down-regulate a deleterious signalling pathway but it may also down-regulate tumour suppressor genes increasing the risk of cancer [40, 41]. This also needs further research and it may prove to be a formidable obstacle in our quest for identifying ideal epigenetic mechanisms for health.

Nevertheless, I will argue that **the conflict for survival between somatic and germ line cells can be conceivably reversed if the survival of somatic cells becomes a priority over the survival of germ line cells.** The mechanism for this is likely to involve hormetic epigenetic influences of the environment upon the soma, and some empirical data supporting this point of view are presented in the next chapter. The effectiveness of this process may be time-dependent, *i.e.* may depend on how early in life the organism is subjected to the 'stressor'. Some evidence exists that if hormetic, epigenetic stressors are applied early in life, these will have a more pronounced and lasting effect [42]. Therefore, although these speculative arguments need more substantial corroboration, it is possible to rely on a small number of emerging studies which may encourage an optimistic outlook.

The Condition-Action Rule

The consequences of adaptation following information capture and subsequent hormetic interference can be better understood through a general principle: The 'Condition-Action' (C-A) rule. Adaptation and thus evolution rely on complex, multilevel applications of several interpretations of 'Condition-Action' rules. The 'Condition-Action' rule is a concept developed in computer science but it can also be applied universally. It may appear simple but carries a remarkable depth of meaning: **in order for an action to be performed, a suitable condition must be met beforehand.** This action may be initiated by a triggering event. A triggering event is a 'significant change in state', a signal that triggers the invocation of the rule [43]. The 'Condition-Action' rule applies to the biology of environment-induced adaptation and to the reallocation of resources from germ line to soma argument:

> **Following a triggering event (*i.e.* the information-carrying hormetic challenge), and if a condition is met (energy resources are required for somatic maintenance-because the arrival of this new information necessitates it), then the action will occur (the resources will be diverted to the soma).**

Here, 'action' is the up-regulation or updates of local data (such as microRNA activity, transposable element actions, epigenetic modifications, exaptation and degeneracy mechanisms *etc.*). Adaptation will occur if the condition triggers an essential event. This C-A rule is executed when the condition becomes true, and it becomes inactive when the condition becomes false. When the condition is true and the C-A rule is executed, there are two possibilities (Fig. **2**):

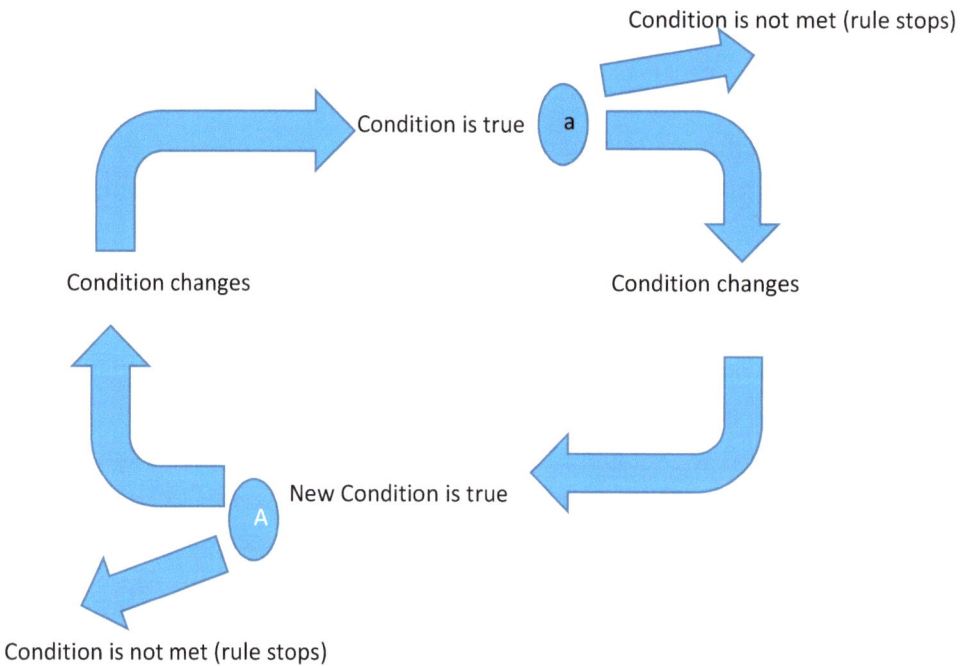

Fig. (2). Schematic flow of a continual invocation of the 'Condition-Action' rule. If the Condition is true, there is Action (a) which leads to modifications in the original condition. If the Condition does not become true, then the rule stops and there is no further action. If the Condition essentially remains true, then there will be further Action (A), and the cycle continues, improving the overall state of the system and making it more robust and adaptable to the new condition with each iteration. For the relevance of this rule to information-dependent reversal the soma-germ line trade-offs, see text above.

A. The condition is executed and after the execution, this condition becomes false (further resources are now not required because there is no further information to assimilate). In this case the C-A rule will not be applied again (and the action will not take place again), or:

B. The condition is executed and the new situation maintains the initial attributes of the condition, making it true again (resources are still required because new information is arriving again). In this case, the rule is activated again (and the new condition is executed). Possibility B is repeated *ad infinitum* until possibility A is invoked.

These are useful concepts to consider during the re-allocation of repair resources from germ to soma, because they provide a standard framework that describes the events and changes that take place. These events and changes obey universal laws which are also found in other domains (such as software development). The general point of this discussion is that, if we continue to be exposed to new and meaningful information, then we may continue to be subjected to the benefits associated with it, *i.e.* continual reallocation of resources to the soma (because somatic cells will require these resources in order to integrate the new information-currying challenges).

Whilst on the subject of the C-A rule it may be worth considering how this applies on a general dichotomy of opinion between gerontologists: whether ageing is stochastic or programmed. If we assume that ageing is programmed then the programme accounts for the elimination of the individual (phenoptosis) in order to improve distribution of resources and information to those more able to survive. In this case therefore the C-A rule would run as follows: If humans are information-poor (the Condition), then there is Action (the elimination of this human and the re-distribution of resources). This leads to modification of the original condition. But if the subsequent condition becomes false (when the human is NOT information-poor because of increased exposure to hormetic information) then the rule stops and there is no further action (there are no further phenoptotic events in this cohort). The programme is not executed. In other words, by deliberately annulling the reason why ageing is programmed, we stop the process from occurring, and ageing stops. Once again, I remind the reader that I define human ageing as **'time-related dysfunction'**.

The Impact of Information upon the Cell

Before I can discuss some additional elements regarding soma to germ line cross-talk (and the effects that this has on lifespan) I will consider the role of information in triggering biological change. Research into the general ways aiming to elucidate how information is captured and used by an organism, has shown that in complex environments where resources are constrained, organisms can evolve integrated brain architectures which have an evolutionary advantage over modular brain structures [44]. In other words, information within a constrained environment, must lead to evolutionary changes that are advantageous for the long term survival of that organism.

Information gathered from the environment helps the organism to improve its fitness and thus its survival. Fig. (1) shows that complexity and fitness as associated. However, this complexity is selected only if it carries some advantage to the organism, and it is not a mere process of aimlessly increasing complexity. The information has to be incorporated within the organism in the physical sense in order to cause a change in a biological status from a less useful to a more useful one. It has been shown that increased fitness is associated with increased capacity to integrate information [45]. It has also been suggested that exposure to complex environments leads to an increased number of options available to the organism which helps integrate the conceptual structures within its brain. Integrated conceptual structures: "*can combine current inputs and outputs with past ones and with the state of internal elements that may reflect past memories as well as future goals*" [44]. Functional complexity in an organism can only be relevant if the environment is also appropriately complex (and, reciprocally, a complex organism helps increase the functional complexity of its own environment). If the environment is simple then organisms do not need much functional complexity. But, given that our environment is becoming more complex, it follows that we too are under increasing pressure to increase our own functional complexity. As mentioned above, information is taken to mean meaningful sets of data which influence the formation or transformation of other data in order to reduce uncertainty (and thus informational entropy) and achieve a goal. In the practical sense, although I acknowledge that the two concepts can be interpreted differently, for the purposes of this discussion I assume that Shannon's information entropy [46] is similar to thermodynamical entropy. Thus my discussion is based upon this general model [47].

Most authorities agree that Environmental Enrichment (EE) (*i.e.* a stimulating and challenging 'information-rich' environment) and hormesis have an effect on several parameters, such as an up-regulation of Brain Derived Neurotrophic Factor (BDNF). In this particular case, it was also shown that an enriched environment early in life results in highly up-regulated histone acetylation at the BDNF gene in adulthood, which happens quite fast after a stimulating exposure [48]. With regards to BDNF, it is also known that its concentration is raised by exercise [49]. The suggestion that a persistent exposure to actionable information forcibly can induce a change in basic, hitherto stable, evolutionary processes is not an entirely speculative one. It has been proposed that increased aerobic activity (thus in this case physical, instead of cognitive, effort) had a direct evolutionary impact on the human brain [50]. The increased and sustained aerobic activity of hunter-gatherers resulted in the up-regulation of peripheral BDNF and increased its concentration to such a degree that it eventually crossed the blood-brain barrier and directly influenced the function and survival of the neurons. The plasma concentrations of other factors that are normally found in the periphery (*i.e.* not in the brain) such as insulin-like growth factor 1 (IGF-1) and vascular endothelial growth factor (VEGF) all of which are elevated during physical activity, were also increased by the sustained physical effort and, having crossed the blood-brain barrier, reached the brain resulting in improved neurogenesis. Thus, it is hypothesized that other sustained activities, in this case cognitive stimulation, may have physical effects which cause a transition from one evolutionary stage to another.

It is known that several epigenetic mechanisms, including methylation of DNA and chromatin remodelling mediate gene regulation in neurons and thus have an important effect in regulating the development of the nervous system. These epigenetic modelling effects influence neural structure and function, and depend upon external environmental clues. Epigenetic gene regulation is important in regulating synaptic plasticity, neural behaviour and high-order cognitive functions such as memory, learning and problem-solving abilities [51].

In order to gain better insights into how information may affect epigenetic regulation, it is useful to explore in some detail the process of information manipulation by the cell. In this case, a relevant example to consider is visual information. When new information is captured by retinal cells there is photonic recycling on the cell surface [52]. When the signal is transmitted to the appropriate neuronal areas (in his case in the occipital cortex) it effects a neuronal response in terms of mainly epigenetic modulation such as DNA acetylation. The steps leading from the acquisition to the assimilation of the incoming information require energy, and this energy needs to be diverted away from other processes in order to be made available to the process that manipulates the information. These adaptation mechanisms depend partly on genetic function but mainly on epigenetic factors, which are able to act fast, and provide more flexibility for optimising performance. Epigenetically-dependent adaptation to visual stimu-lation can be very rapid [53]. Thus, the environment, through epigenetic regulation affects brain function, and it follows that if environmental stimulation is of a certain magnitude and quality, it will have direct effects on cognition. Cognition and higher mental functions are essential in our current technological, high information-dependant environment, and those who are able to adapt to these challenges are more likely to survive and be healthy within this milieu.

Small Non-Coding RNAs

Recent research has shed further light in some of the mechanisms of the effects of the environment upon epigenetic regulation. For instance, it is known that genetic elements such as microRNAs (miRNAs) are effective epigenetic regulators of a genomic response to environmental stimuli.

Environmental agents such as O3 (ozone) exposure can disrupt miRNA expression profiles and interfere with inflammation regulation [54]. MicroRNAs contribute to genome plasticity and evolution, and are essential in controlling the expression of genes which may interfere with the ageing process. However, further study is needed in this respect, as many effects of miRNA regulation are still unknown [55]. What is known is that the number and function of miRNAs are significantly up-regulated in humans compared to other animals such as chimpanzee, macaque, mouse, rat or dog [56]. The authors [57] quote:

Gene expression regulation is a complex and highly organized process involving a variety of genomic factors. It is widely accepted that differences in gene expression can contribute to the phenotypic variability between species, and that their interpretation can aid in the understanding of the physiologic variability. CNVs (copy number variations) and miRNAs are two major players in the regulation of expression plasticity and may be responsible for the unique phenotypic characteristics observed in different lineages. We have previously demonstrated that a close interaction between these two genomic elements may have contributed to the regulation of gene expression during evolution.... A comprehensive analysis of these interactions indicates a unique nature of human CNV genes regulation as compared to other species. ...We demonstrate a distinct and tight regulation of human genes that might explain some of the unique features of human physiology. In addition, comparison of gene expression regulation between species indicated that there is a significant difference between humans and mice possibly questioning the effectiveness of the latest as experimental models of human diseases

Here the authors highlight the fact that humans may be quite unique with respect to gene regulation *via* environmental influences, and also that extrapolating from research in mice to clinical benefits in humans may be an invalid concept (which, by the way, shows the futility of rejuvenation biotechnological research, which is largely based on laboratory inferences).

The fact that humans exhibit a rather distinctive microRNA profile may indicate that we may have developed unique evolutionary genetic elements that can facilitate a better repair of damage, exactly because of the need to adapt quickly to an information-rich technological environment [58], an evolutionary scenario which is not relevant in other non-human animals.

There are several other ways by which microRNAs translate external environmental cues into biologically-usable information. The role of microRNA in Alzheimer's disease has been studied in some detail [59]. It is known that an information-rich environment can slow the progression of the disease but the detailed mechanisms of this are still being studied. In this particular study, EE exhibited an effect on post-transcriptional regulating factors and microRNAs. These authors reported that miRNAs (particularly synaptic microRNAs) were down-regulated in Alzheimer's dementia and subsequently up-regulated following exposure to an enriched environment. In addition, the authors found that tomosyn (an inhibitor of synaptic transmission) is regulated by the function of microRNAs (which block tomosyn in Alzheimer's dementia following EE). These results show how an enriched environment can affect biological agents such as microRNAs, which have specific and well-orchestrated effects downstream, with a resulting functional improvement or decrease risk of some age-related disease. Therefore, micro RNAs are heavily involved in translating environmental information into biological change. For instance, EE down-regulates the expression of the glucocorticoid receptor through the mediation of the microRNA-124a [60] and this may subsequently reduce anxiety and improve behaviour. Thus, there exists an increasingly-studied signalling pathway providing the basis for explaining how external information affects our biological actions. It is likely that other mechanisms could be involved, although the importance of microRNA remains high.

Finally, it is worth mentioning that a chronically unpredictable, but intrinsically safe environment (*i.e.* as our constantly changing society and techno-culture) can induce phenotypic changes in the parents, which can be reflected in the unexposed offspring, suggesting that at least some of the epigenetic changes can be heritable. This adds robustness to the notion suggesting that epigenetic changes are important in evolution, as these are more permanent and can be subjected to evolutionary selection. Furthermore, repeated stress early in life can protect against more sustained stress later in life [61] and social factors can play a role in epigenetic regulation, the effects of which not only persist throughout life, but can also be transmitted to the offspring *via* epigenetic inheritance [62]. In addition, there are specific epigenetic profiles associated with longevity [63] which can be transmitted to the offspring. Gentilini *et al.* [64] have shown that centenarians have certain specific DNA methylation characteristics (such as those found in nucleotide biosynthesis, and control of signal transmission) which are also found in their offspring, suggesting that these characteristics are heritable.

CONCLUSION

In summary, there are several mechanisms, some well-described, and some still to be explained, that may cause epigenetic changes following exposure to actionable information. External challenges such as hormetic stimulation and information-that-requires-action, encountered within a technological society, have clear effects on our biology and may play a part in shifting the emphasis of resource allocation for repairs, from the germ line to the soma. This is a crucial process that can explain why hormesis may have effects which are well beyond what is currently recognised, effects which do not only improve general health but may also fundamentally change our biology and evolution. This concept is further discussed in the following chapter.

REFERENCES

[1]   Price TD, Qvarnström A, Irwin DE. The role of phenotypic plasticity in driving genetic evolution. Proc Biol Sci. 270(1523): 1433-40.
       [http://dx.doi.org/10.1098/rspb.2003.2372]

[2]   Kelly SA, Panhuis TM, Stoehr AM. Phenotypic plasticity: molecular mechanisms and adaptive significance. Compr Physiol 2012; 2(2): 1417-39.
       [PMID: 23798305]

[3]   Takano T, Nakamura K, Watanabe M. Urban residential environments and senior citizens' longevity in megacity areas: the importance of walkable green spaces. J Epidemiol Community Health 2002; 56(12): 913-8.
       [http://dx.doi.org/10.1136/jech.56.12.913] [PMID: 12461111]

[4]   Frieden BR, Gatenby RA. Information dynamics in living systems: prokaryotes, eukaryotes, and cancer. PLoS One 2011; 6(7): e22085.
       [http://dx.doi.org/10.1371/journal.pone.0022085] [PMID: 21818295]

[5]   Kyriazis M. Reversal of informational entropy and the acquisition of germ-like immortality by somatic cells. Curr Aging Sci 2014; 7(1): 9-16.
       [http://dx.doi.org/10.2174/1874609807666140521101102] [PMID: 24852017]

[6]   Cohen A. Complex systems dynamics in aging: new evidence, continuing questions Biogerontology. Available from: http://link.springer.com/article/10.1007/s10522-015-9584-x 2015.

[7]   Edlund JA, Chaumont N, Hintze A, Koch C, Tononi G, Adami C. Integrated information increases with fitness in the evolution of animats. PLOS Comput Biol 2011; 7(10): e1002236.
       [http://dx.doi.org/10.1371/journal.pcbi.1002236] [PMID: 22028639]

[8]   Ashby WR. An introduction to cybernetics. London: Chapman & Hall 1956.
       [http://dx.doi.org/10.5962/bhl.title.5851]

[9]   Smart J. Measuring Innovation in an Accelerating World. Technol Forecast Soc Change 2005; 72(8): 988-95.
       [http://dx.doi.org/10.1016/j.techfore.2005.07.001]

[10]   Heylighen F. Fitness as Default: the evolutionary basis for cognitive complexity reduction. In: Trappl R, Ed. Cybernetics and Systems. Singapore: World Science 1994; pp. 1595-602.

[11]   Gravina S, Vijg J. Epigenetic factors in aging and longevity. Pflugers Arch 2010; 459(2): 247-58.
       [http://dx.doi.org/10.1007/s00424-009-0730-7] [PMID: 19768466]

[12]   Combes AN, Whitelaw E. Epigenetic reprogramming: enforcer or enabler of developmental fate? Dev Growth Differ 2010; 52(6): 483-91.
       [http://dx.doi.org/10.1111/j.1440-169X.2010.01185.x] [PMID: 20608951]

[13]   Carnes BA. What is lifespan regulation and why does it exist? Biogerontology 2011; 12(4): 367-74.
       [http://dx.doi.org/10.1007/s10522-011-9338-3] [PMID: 21512719]

[14]   Walker RF. Developmental theory of aging revisited: focus on causal and mechanistic links between development and senescence. Rejuvenation Res 2011; 14(4): 429-36.
       [http://dx.doi.org/10.1089/rej.2011.1162] [PMID: 21767161]

[15]   Takesian AE, Hensch TK. Balancing plasticity/stability across brain development. Prog Brain Res 2013; 207: 3-34.
       [http://dx.doi.org/10.1016/B978-0-444-63327-9.00001-1] [PMID: 24309249]

[16]   Huang S. The molecular and mathematical basis of Waddington's epigenetic landscape: a framework for post-Darwinian biology? BioEssays 2012; 34(2): 149-57.
       [http://dx.doi.org/10.1002/bies.201100031] [PMID: 22102361]

[17]   Haggarty P, Hoad G, Harris SE, *et al.* Human intelligence and polymorphisms in the DNA methyltransferase genes involved in epigenetic marking. PLoS One 2010; 5(6): e11329.
       [http://dx.doi.org/10.1371/journal.pone.0011329] [PMID: 20593030]

[18]   Santos F, Dean W. Epigenetic reprogramming during early development in mammals. Reproduction 2004; 127(6): 643-51.
       [http://dx.doi.org/10.1530/rep.1.00221] [PMID: 15175501]

[19]   Vaiserman AM. Hormesis and epigenetics: is there a link? Ageing Res Rev 2011; 10(4): 413-21.
       [PMID: 21292042]

[20]   Jablonka E, Lamb MJ. Epigenetic inheritance in evolution. J Evol Biol 1998; 11(2): 159-83.
       [http://dx.doi.org/10.1007/s000360050073]

[21]   Calabrese EJ, Mattson MP. Hormesis provides a generalized quantitative estimate of biological plasticity. J Cell Commun Signal 2011; 5(1): 25-38.
       [http://dx.doi.org/10.1007/s12079-011-0119-1] [PMID: 21484586]

[22]   Rattan SI. Molecular gerontology: from homeodynamics to hormesis. Curr Pharm Des 2014; 20(18): 3036-9.
       [http://dx.doi.org/10.2174/13816128113196660708] [PMID: 24079765]

[23]   Calabrese V, Cornelius C, Dinkova-Kostova AT, Calabrese EJ, Mattson MP. Cellular stress responses, the hormesis paradigm, and vitagenes: novel targets for therapeutic intervention in neurodegenerative disorders. Antioxid Redox Signal 2010; 13(11): 1763-811.
[http://dx.doi.org/10.1089/ars.2009.3074] [PMID: 20446769]

[24]   Mattson MP. Awareness of hormesis will enhance future research in basic and applied neuroscience. Crit Rev Toxicol 2008; 38(7): 633-9.
[http://dx.doi.org/10.1080/10408440802026406] [PMID: 18709572]

[25]   Agutter PS. Elucidating the Mechanism(s) of Hormesis at the Cellular Level: The Universal Cell Response. Am J Pharmacol Toxicol 2008; 3(1): 100-10.
[http://dx.doi.org/10.3844/ajptsp.2008.100.110]

[26]   Vaiserman AM. Hormesis, adaptive epigenetic reorganization, and implications for human health and longevity. Dose Response 2010; 8(1): 16-21.
[http://dx.doi.org/10.2203/dose-response.09-014.Vaiserman] [PMID: 20221294]

[27]   Vaiserman AM. Epigenetic engineering and its possible role in anti-aging intervention. Rejuvenation Res 2008; 11(1): 39-42.
[http://dx.doi.org/10.1089/rej.2007.0579] [PMID: 18260779]

[28]   Fraga MF, Ballestar E, Paz MF, *et al.* Epigenetic differences arise during the lifetime of monozygotic twins. Proc Natl Acad Sci USA 2005; 102(30): 10604-9.
[http://dx.doi.org/10.1073/pnas.0500398102] [PMID: 16009939]

[29]   Heylighen F. Cybernetic principles of aging and rejuvenation: the buffering- challenging strategy for life extension. Curr Aging Sci 2014; 7(1): 60-75.
[http://dx.doi.org/10.2174/1874609807666140521095925] [PMID: 24852018]

[30]   Westendorp RG, Kirkwood TB. Human longevity at the cost of reproductive success. Nature 1998; 396(6713): 743-6.
[http://dx.doi.org/10.1038/25519] [PMID: 9874369]

[31]   Heininger K. Aging is a deprivation syndrome driven by a germ-soma conflict. Ageing Res Rev 2002; 1(3): 481-536.
[http://dx.doi.org/10.1016/S1568-1637(02)00015-6] [PMID: 12067599]

[32]   Cossetti C, Lugini L, Astrologo L, Saggio I, Fais S, Spadafora C. Soma-to-germline transmission of RNA in mice xenografted with human tumour cells: possible transport by exosomes. PLoS One 2014; 9(7): e101629.
[http://dx.doi.org/10.1371/journal.pone.0101629] [PMID: 24992257]

[33]   Baudisch A, Vaupel JW. Evolution. Getting to the root of aging. Science 2012; 338(6107): 618-9.
[http://dx.doi.org/10.1126/science.1226467] [PMID: 23118175]

[34]   Kirkwood TB. Immortality of the germ-line *versus* disposability of the soma. Basic Life Sci 1987; 42: 209-18.
[PMID: 3435387]

[35]   Smelick C, Ahmed S. Achieving immortality in the *C. elegans* germline. Ageing Res Rev 2005; 4(1): 67-82.
[http://dx.doi.org/10.1016/j.arr.2004.09.002] [PMID: 15619471]

[36]   Gracida X, Eckmann CR. Fertility and germline stem cell maintenance under different diets requires nhr-114/HNF4 in C. elegans. Curr Biol 2013; 23(7): 607-13.
       [http://dx.doi.org/10.1016/j.cub.2013.02.034] [PMID: 23499532]

[37]   Gracida X, Eckmann CR. Mind the gut: Dietary impact on germline stem cells and fertility. Commun Integr Biol 2013; 6(6): e26004.
       [http://dx.doi.org/10.4161/cib.26004] [PMID: 24563704]

[38]   Avise JC. The evolutionary biology of aging, sexual reproduction and DNA repair. Evolution 1993; 47: 1293-301.
       [http://dx.doi.org/10.2307/2410148]

[39]   Santos F, Dean W. Epigenetic reprogramming during early development in mammals. Reproduction 2004; 127(6): 643-51.
       [http://dx.doi.org/10.1530/rep.1.00221] [PMID: 15175501]

[40]   Kemp CJ, Moore JM, Moser R, *et al.* CTCF haploinsufficiency destabilizes DNA methylation and predisposes to cancer. Cell Reports 2014; 7(4): 1020-9.
       [http://dx.doi.org/10.1016/j.celrep.2014.04.004] [PMID: 24794443]

[41]   Choi JD, Lee JS. Interplay between epigenetics and genetics in cancer. Genomics Inform 2013; 11(4): 164-73.
       [http://dx.doi.org/10.5808/GI.2013.11.4.164] [PMID: 24465226]

[42]   Burger JM, Hwangbo DS, Corby-Harris V, Promislow DE. The functional costs and benefits of dietary restriction in *Drosophila.* Aging Cell 2007; 6(1): 63-71.
       [http://dx.doi.org/10.1111/j.1474-9726.2006.00261.x] [PMID: 17266676]

[43]   Mani Chandy K. Event-Driven Applications: Costs, Benefits and Design Approaches. California Institute of Technology 2006.

[44]   Albantakis L, Hintze A, Koch C, Adami C, Tononi G. Evolution of integrated causal structures in animats exposed to environments of increasing complexity. PLOS Comput Biol 2014; 10(12): e1003966.
       [http://dx.doi.org/10.1371/journal.pcbi.1003966] [PMID: 25521484]

[45]   Joshi NJ, Tononi G, Koch C. The minimal complexity of adapting agents increases with fitness. PLOS Comput Biol 2013; 9(7): e1003111. [46].
       [http://dx.doi.org/10.1371/journal.pcbi.1003111] [PMID: 23874168]

[46]   Shannon CE. A Mathematical Theory of Communication. Bell Syst Tech J 1948; 27(3): 379-423.
       [http://dx.doi.org/10.1002/j.1538-7305.1948.tb01338.x]

[47]   Heylighen F, Joslyn C. Cybernetics and second order cybernetics. Encyclopedia of Physical Science & Technology. 3. New York: Academic Press 2001; 4: pp. 155-70.

[48]   Branchi I, Karpova NN, D'Andrea I, Castrén E, Alleva E. Epigenetic modifications induced by early enrichment are associated with changes in timing of induction of BDNF expression. Neurosci Lett 2011; 495(3): 168-72.
       [http://dx.doi.org/10.1016/j.neulet.2011.03.038] [PMID: 21420470]

[49]   Ferris LT, Williams JS, Shen CL. The effect of acute exercise on serum brain-derived neurotrophic factor levels and cognitive function. Med Sci Sports Exerc 2007; 39(4): 728-34.

[http://dx.doi.org/10.1249/mss.0b013e31802f04c7] [PMID: 17414812]

[50] Raichlen DA, Polk JD. Linking brains and brawn: exercise and the evolution of human neurobiology. Proc Biol Sci. 280(1750)

[51] Feng J, Fouse S, Fan G. Epigenetic regulation of neural gene expression and neuronal function. Pediatr Res 2007; 61(5 Pt 2): 58R-63R.
[http://dx.doi.org/10.1203/pdr.0b013e3180457635] [PMID: 17413844]

[52] Saxena A, Jacobson J, Yamanashi W, Scherlag B, Lamberth J, Saxena B. A hypothetical mathematical construct explaining the mechanism of biological amplification in an experimental model utilizing picoTesla (PT) electromagnetic fields. Med Hypotheses 2003; 60(6): 821-39.
[http://dx.doi.org/10.1016/S0306-9877(03)00011-2] [PMID: 12699707]

[53] Adorjan P, Schwabe L, Wenning G, Obermayer K. Rapid adaptation to internal states as a coding strategy in visual cortex? Neuroreport 2002; 13(3): 337-42.
[http://dx.doi.org/10.1097/00001756-200203040-00018] [PMID: 11930134]

[54] Fry RC, Rager JE, Bauer R, *et al.* Air toxics and epigenetic effects: ozone altered microRNAs in the sputum of human subjects. Am J Physiol Lung Cell Mol Physiol 2014; 306(12): L1129-37.
[http://dx.doi.org/10.1152/ajplung.00348.2013] [PMID: 24771714]

[55] Le TD, Liu L, Zhang J, Liu B, Li J. From miRNA regulation to miRNA-TF co-regulation: computational approaches and challenges. Brief Bioinform 2015; 16(3): 475-96.
[http://dx.doi.org/10.1093/bib/bbu023] [PMID: 25016381]

[56] Dweep H, Georgiou GD, Gretz N, Deltas C, Voskarides K, Felekkis K. CNVs-microRNAs interactions demonstrate unique characteristics in the human genome. An interspecies in silico analysis. PLoS One 2013; 8(12): e81204.
[http://dx.doi.org/10.1371/journal.pone.0081204] [PMID: 24312536]

[57] Dweep H, Kubikova N, Gretz N, Voskarides K, Felekkis K. Homo sapiens exhibit a distinct pattern of CNV genes regulation: an important role of miRNAs and SNPs in expression plasticity. Sci Rep 2015; 5: 12163.
[http://dx.doi.org/10.1038/srep12163] [PMID: 26178010]

[58] Heylighen F. Evolutionary psychology. In: Michalos A, Ed. Encyclopedia of Quality of Life Research. Berlin: Springer 2011.

[59] Barak B, Shvarts-Serebro I, Modai S, *et al.* Opposing actions of environmental enrichment and Alzheimer's disease on the expression of hippocampal microRNAs in mouse models. Transl Psychiatry 2013; 3: e304.
[http://dx.doi.org/10.1038/tp.2013.77] [PMID: 24022509]

[60] Durairaj RV, Koilmani ER. Environmental enrichment modulates glucocorticoid receptor expression and reduces anxiety in Indian field male mouse Mus booduga through up-regulation of microRNA-124a. Gen Comp Endocrinol 2014; 199: 26-32.
[http://dx.doi.org/10.1016/j.ygcen.2014.01.005] [PMID: 24457250]

[61] Natt D. Heritable epigenetic changes to environmental challenges 2011.

[62] Thayer ZM, Kuzawa CW. Biological memories of past environments: epigenetic pathways to health disparities. Epigenetics 2011; 6(7): 798-803.

[http://dx.doi.org/10.4161/epi.6.7.16222] [PMID: 21597338]

[63]   Gentilini D, Castaldi D, Mari D, *et al.* Age-dependent skewing of X chromosome inactivation appears delayed in centenarians' offspring. Is there a role for allelic imbalance in healthy aging and longevity? Aging Cell 2012; 11(2): 277-83.
[http://dx.doi.org/10.1111/j.1474-9726.2012.00790.x] [PMID: 22292741]

[64]   Gentilini D, Mari D, Castaldi D, *et al.* Role of epigenetics in human aging and longevity: genome-wide DNA methylation profile in centenarians and centenarians' offspring. Age (Dordr) 2013; 35(5): 1961-73.
[http://dx.doi.org/10.1007/s11357-012-9463-1] [PMID: 22923132]

CHAPTER 5

A 'War of Trade-offs' Between the Soma and the Germ Line

Abstract: In nature there is always a fierce competition for resources. This is particularly relevant during the process of ageing, where there is canalisation of repair resources; these tend to flow from the somatic tissue towards the germ line, in order to assure the survival of the species. In the early periods of phylogenetic development there was a time when repair of germ line cells became more efficient compared to the repair of somatic cells. The level of somatic repair became just sufficient to ensure that the organism reached sexual maturity. We are now looking for evidence that this could be changing, that we may be witnessing a **phase transition** from effective germ line repair to an effective somatic repair. Here, I consider mechanisms of fidelity-preservation which may be present in the germ line, and examine the possibility that these may be made to operate upon somatic cells instead. Mechanisms which safeguard the reliability of germ line repair and ensure robustness/resilience in the germ line may also (or instead) be applicable upon somatic material (cells, molecules, and other factors) and safeguard a continually-effective repair of this somatic material. Relentless hormetic challenges from the environment guarantee that the flow of information remains operational and it persistently fuels the ability to repair the soma. Apart from germ line cells, some unicellular organisms such as certain bacteria maintain their ability for ongoing repairs (at least for some considerable time, albeit not indefinitely), a fact that indicates that, in principle, senescence is not unavoidable. Thus the ability to repair and maintain somatic organic material within biological systems is not entirely lost.

Keywords: Apoptosis, Environment, Germ line, Germ line to soma cross-talk, Indispensable soma hypothesis, Immortalisation, MicroRNAs, Somatic repairs, Trade-offs, Transposons.

Marios Kyriazis

INTRODUCTION

In this chapter, I will discuss, review and speculate on matters relating to how exposure to information (in humans) may lead to a situation whereby ageing as a process may be significantly downgraded (or even virtually eliminated, at least in some sections of humanity). This is particularly relevant if we accept that ageing is programmed, when we could conceivably stop the programme as mentioned in the Condition-Action rule previously. However, the discussion is also relevant if ageing is stochastic, when random damage causes dysfunction. This is because the discussion accounts for repair of age-related damage, even when this damage is random. Some initial concepts have already been discussed in the previous chapter but I will continue addressing this matter from a variety of angles. In this chapter and throughout the book, I define human ageing simply as 'time-related dysfunction'. This accounts for the fact that ageing is a process dependent on the passage of time, and it is only of interest to humans because it results in loss of function (and consequently, chronic degenerative disease and death). This concept is important because it places the discussion, not only on a biological basis but also on a clinical, everyday life basis.

This argument, and most of the concepts found in this book, is based upon certain simple evolutionary principles. It is indisputable that the widespread tendency in nature is towards continual survival and good function. When a **mildly** stressful (*i.e.* not excessive or prolonged) event happens, such as famine or starvation, the reproductive priorities in humans are down-graded [1]. The crucial question to ask here is 'why does this happen?' Again, it is undeniable, in my view, that the answer is this: When life is immediately in danger, there is a 'hard-wired' tendency in natural principles to protect the organism as a priority, and allocate to this process whatever resources are necessary for effective repair and maintenance. This may mean a reduced priority for reproductive resources. There are species that will hold off on reproduction, or send embryos into diapause, during hard times, such as the honey bee, or the roe deer for instance. What is relevant here is that, although this does not mean that the organism lives longer, at least there is strong evidence that organisms can reallocate resources from reproduction to tissue maintenance. There exist definite trade-offs between somatic maintenance and reproduction [2], but when somatic elements need repair resources, the first initial objective (*i.e.* the 'default' option) is to allocate these to the soma. If the danger continues, or if the environment is risky and there is an increased likelihood of the organism dying early, then nature switches to 'option B', the second best option for ensuring survival, which involves the withholding of somatic repair resources and the reallocation of these to the germ line [3]. The point of this argument is to show that the tendency to survive as a discrete organism is **innate and present now** in all of us, and it is the first priority of most biological processes. Ageing and reproduction are merely secondary processes 'developed' by natural principles in order to assure survival, only as an ancillary, reserve mechanism. The above concepts should be taken within the definition I give to the term 'evolution'. For me, <u>evolution means 'the adaptation to changes in the environment, in order to continue survival'.</u>

In the co-evolution of the repair mechanisms employed by somatic and germ line cells, there was a certain antagonism, whereby germ line elements have succeeded in modifying the opponent's (somatic) control systems (for instance [4]). Despite several countermeasures deployed by somatic cells in order to acquire sufficient repair resources, there was a relatively rapid divergence of the functionality of the control and regulation systems, resulting in the immortality of the germ line with the mortality of the soma [5]. It is possible to study the basic theoretical mechanisms involved in such co-evolutionary setting, and consider ways to modify or interfere with the processes, in a way that favours the soma instead of the germ line. This may lead to methods which study a fitness landscape where somatic repair mechanisms can evolve rapidly [6]. In this respect, Smelick and Ahmed [7] have suggested that the germ line can antagonise the ageing of somatic cells, and that it may be possible that defects in the mechanisms operating during immortalisation of germ line cells may provide useful repair resources to somatic cells. Germ cells achieve continuous repair and fidelity of replication by ensuring that they maintain robustness - the redundancy that counteracts the effects of random damage. Evolution drives the balance of the appropriate trade-offs between robustness and maintenance resources. The trade-offs between survival of the somatic cells and reproduction could be due to factors such as:

a. The impossibility to maintain all processes within the body indefinitely, due to lack of repair resources (the notion of 'Clashing demands' in physiology: because many properties of the metabolism are closely interrelated, it is often possible to keep one property relatively stable only by moving another property away from its normal steady state. This has repercussions on this second property which has been moved away from its steady state, but the overall result is that the organism can function effectively again).

b. The damage to the somatic cells could be incurred during the normal course of reproduction [8].

Germ-Line Replicative Fidelity

Under conditions of continual, excessive stress, repair resources are preferentially allocated from somatic to germ line repair in order to safeguard the health of the next generation (see Box **1**) [9]. This is essentially the basis of the Disposable Soma theory [3]. Until recently, it was believed that this process is unidirectional, with resources flowing from soma to germ line only. There is evidence however supporting the suggestion that this process is, in fact, *bi-directional*. Under suitable conditions, repair resources can be re-directed from the germ line back to the somatic tissues [10]. This germ-initiated somatic protective response is called 'germ-line DNA-damage-induced systemic stress resistance (GDISR).

BOX 1. What is a 'somatic' cell

It is important to emphasise that in the following discussion (and throughout the book) by 'somatic cell' or 'soma' I essentially refer to a **neuron**, or to any other relevant cell or organic agent that carries, stores, transmits or elaborates information, AND ALSO to all other cells or tissues which are necessary for the survival of this primary neuron. In essence then:

A. A somatic cell is a neuron plus any other non-germ cell that support the neuron's function, proximally or distantly.
B. A germ line cell is a reproductive egg or sperm cell, with the possible inclusion of reproductive stem cells such as spermatogonial and ovarian germ line stem cells.

There is a direct relationship between increasing age and genomic instability in somatic cells. We know that allocation of maintenance resources favours reproduction at the expense of somatic cell senescence. However, it is possible to encounter soma-to-germ line transformation of gene expression, such as the improved function of the transcription factor Daf-16, which is normally encountered only in the germ line [11]. There is a substantial body of evidence showing that the rejuvenation process encountered in germ line erases the age-related damage that accumulates over the years. Three mechanisms have been suggested to account for the functional stability of the genome in germ line [7]:

1. A generally more efficient and increased rate of cellular repair and maintenance, as well as specific repair and rejuvenation mechanisms. It is known for example, that DNA repair systems including GG-NER (Global Genome Nucleotide Excision Repair) are specifically active in germ line cells and contribute towards the effective repair of the germ DNA, but these are also active in somatic cells [12]
2. Efficient selection of fully functional germ cells which are allowed to propagate at the expense of less efficient germ cells [13]. A relevant concept here is that of apoptosis regulation, which may also depend on neuronal inputs as will be discussed below.
3. Non-autonomous contributions of the soma to germ line rejuvenation [14]. This may be one of the reasons why somatic stem cells become less numerous and less functional with age: due to repair resources being continually diverted to the germ line, there is an overall inability to maintain good function in somatic stem cells, which also lose their ability to differentiate properly.

As hinted above, such strategies may also be present in somatic cells, but are significantly down-regulated. It is tempting to examine the possibility that this rejuvenation process can be reversed, and driven to operate effectively in somatic cells. It is important to highlight that certain mechanisms of germ line rejuvenation could be dependent upon epigenetic modifications and factors that regulate transcription. Thus it is also conceivable that careful epigenetic co-ordination may have certain rejuvenating effects upon somatic cells, as discussed in the previous chapter.

The above facts raise the possibility that somatic cells lines may, under certain circumstances, withhold any 'immortality' contributions for their own repair instead. This may happen when somatic cells are under intense pressure to maintain themselves (following, for example, unremitting information inputs and challenges). Fontana [15] has suggested that when 'good' (*i.e.* immortality-affording) matching sequences are not present in the germ line genome, then these 'good' sequences are created in the somatic cell and subsequently migrate to the germ line through the bloodstream, a process he termed Germ Line Penetration. In this case, he suggests that transposition of genomic elements in somatic cells drives differentiation in germ cells that drive evolution. As a consequence, it can also be argued that this process can conceivably be arrested, with somatic cells retaining the immortality-affording sequences and using these for their own repairs instead, <u>if</u> this is beneficial to the overall evolutionary process.

Recent findings that neurons may influence germ line survival lends support to this line of thinking. Levi-Ferber *et al.* [16] have shown that neuronal stress induces apoptosis in the germ line. This process is mediated by the IRE-1 factor, an endoplasmic reticulum stress response sensor, which then activates p53 and initiates the apoptotic cascade in the germ line. Phosphorylated IRE-1 also activates tumour-necrosis factor (TNF)-receptor-associated factor 2 (TRAF2) which is another apoptosis-initiating factor [17]. Therefore, germ cells may die due to a stress response originating from the distantly-located neurons. The process of apoptosis depends on a large number of converging inputs but it is significant that there is an apparent conflict between neurons and germ cells. If this is confirmed to be the case, it will become a crucial supporting element and will help the entire thrust of the arguments promulgated in this book: **that cognitive stimulation, initiates stress response in neurons which then may effect apoptosis of (or other damage in) germ line cells thus ensuring their own continual survival and function, changing the current patterns of human biological evolution** (Fig. **1**).

On the opposite side of the coin, it was recently shown that germ line cells 'fight back' and increase degeneration of the neurons. Initially, researchers Wu *et al.* [18] were studying the effects of pathogen infection upon the initiation of neuronal degeneration in *C.elegans*.

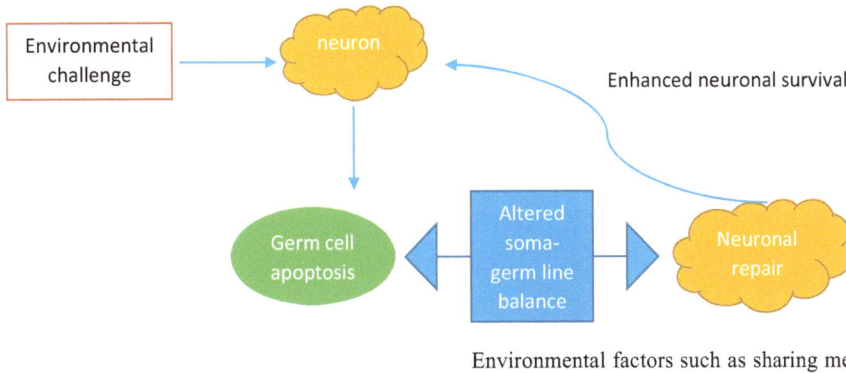

Environmental factors such as sharing meaningful

Fig. (1). How information ensures neuronal survival. Environmental factors such as sharing meaningful information-that-requires-action, impact positively on the neuron, which may initiate a stress response and then germ cell apoptosis. This changes the balance between germ line *vs* somatic repair resource allocations. The result is that neurons are better maintained and continue to survive in tandem with diminished germ cell survival. The energetic balance is maintained and the compliance with the second law of thermodynamics remains intact.

Their hypothesis was that the immune response following infection results in neurodegeneration. But they also found that germ line loss resulted in resistance to neurodegeneration following infection. In other words, germ line factors facilitate degeneration of the neurons, and when these germ line factors are lost, then neural degeneration decreases. These effects are mediated by the maternal sterile gene *mes-1* which encodes a receptor protein that is needed for the essential (from the germ line point of view) unequal cell division in embryonic germ line stem cells. This division is required in order for gamete selection, to segregate mutation-carrying material and eliminate it so that it will not interfere with the future health of the germ line cells. The authors found that the gene *pgl-1* which is required for healthy germ line cell development, and it was also affected. The possibility exists that germ cells may out-compete the neuronal cells for a specific resource, such as essential fatty acids or other factors. If this is shown to be the case than it would strengthen the case that there is direct action of germ cells upon the neurons. It may also be that if such a link is proven, then it could be possible to interfere with this competition and reverse the flow of resources so that to favour the neuron.

The significance of these findings is staggering. It is as if neurons are targeting the germ line and aim to damage it (through apoptosis and the IRE-1 example mentioned above), but the germ line deploys countermeasures targeting neurons and aims to destroy **them**. Are we witnessing a kind of war of trade-offs between the germ line and the neurons? It is tempting to answer 'yes', although these are results that need confirming. This speculation is intriguing and, if true, it may prove a crucial mechanism which we may try to manipulate - I posit that we can indeed manipulate it through information that causes neuronal up-regulation. The results of this manipulation could be the rebalancing of trade-offs between soma and germ line (and thus the reduction of age-related dysfunction).

Soma-to-Germ Line Communication

As a general principle, it can be assumed that exposure to an information-rich environment results in an increase in the information content of the individual. This information impacts on (and may influence) biological processes as discussed above. It is known that one of the ways information can be transmitted from the environment to the genome (and, reciprocally, from the genome to the phenotype and thus back to the environment) is through microRNAs. This has been discussed in the previous chapter and will be discussed again here.

Any suitable environmental challenge can cause epigenetic effects which can influence inheritance of certain features through microRNA mediation. This characteristic is encountered in plants, worms and also mammals [19]. The study of "transgenerational systems biology" is concerned with identifying elements involved in heritable epigenetic changes, and it can also provide us with valuable clues about the behaviour and properties of information transfer between the soma and the germ line [20]. However, the suggestion that there is transfer of heritable information from soma to germ line is against conventional and established knowledge, confounded by the fact that epigenetic memory needs to be maintained throughout the process. Nevertheless, research is now providing answers to these conundrums. Sharma [21] has shown that new evidence now supports the transfer of heritable epigenetic information from soma to germ line and highlights the role of microRNAs and exosome during this process.

Several epigenetic factors have been implicated in germ line transmission of information, and, although some details are discussed below, the full details regarding the mediators of this information transfer from soma to the germ line are still poorly understood. Germ cells are resistant to age-related degeneration, and although I have discussed some general principles above, the exact mechanisms of this resistance are not yet fully known. What is known is that one germ line-specific mechanisms depends on microRNAs and PIWI (P-element Induced WImpy testis) proteins that regulate the function of transposons, resulting in an efficient repair of germ line damage [22]. Therefore, extracellular microRNAs (miRNAs) are involved in this epigenetic inheritance in mammals [23]. With regards to the role of PIWIs in germ line immortality, there is a balance between transposon action (which inhibits germ line cell differentiation) and PIWI action which silences the expression of genetic elements, maintaining an optimal function of the germ cell [24]. MicroRNAs are regulated by other non-coding RNAs such as the intergenic RNA. An imbalance of this regulation can lead to disease and dysfunction [25]. This helps us examine and understand better the overall function and value of microRNAs. It increasingly appears more likely that non-coding RNAs may mediate transfer of information from somatic cells to germ cells. Further evidence suggests that exosomal microRNAs and exosomal proteins may also mediate information transfer from soma to germ line [23]. This is important because it suggests that environmental factors can induce epigenetic adaptations, and alter the information transfer from soma to germ line, with results which can be experienced not only in individual organisms but can also be felt through generations. Siblings produced from young or ageing parents have similar lifespans, and this further highlights the fact that germ line cells are resistant to age-related degeneration. Of course, there are many other factors that afford fidelity of information transfer in germ cells.

Apart from microRNAs there are other mechanisms involved in soma to germ line cross-talk. Certain interventions which downgrade reproduction may also cause a lifespan extension. Ablation of germ cells in *C. elegans* leads to an increased lifespan which shows that signals from the germ line have a direct impact upon somatic cell survival, and this may be due to an increased resistance of somatic cells to stress [11]. One mechanism involves the FOXO-family transcription factors (such as Daf-16) in somatic tissues, which are up-regulated following signals from damaged germ line cells [26]. Daf-16 then regulates downstream genes which are involved in somatic life-span extension, at least in *C.elegans*. Additional loss of germ line cells, increases further the already long lifespan of somatic cells [26].

Intracellular clearance systems are also up-regulated following signals from the germ line [27]. This is significant because it indicates that germ cells have direct control over the health and longevity of somatic cells. In addition to this, protein homoeostasis in somatic cells is well-maintained when germ cells are damaged and it is significantly downgraded when germ cell function increases [28]. There exist mechanisms in germ cells which may induce somatic cell reprogramming and somatic stem cell pluripotency [29, 30]. Kim *et al.* [31] have shown that when male sticklebacks experience a benign and safe environment:

> "*...subsequently reduced their investment in carotenoid-based sexual signals early in the breeding season, and consequently**senesced at a slower rate**later in the season, compared to those that had developed under harsher conditions. This plasticity of ageing was genetically determined. Both antagonistic pleiotropy and genetic variation in the rate of senescence were evident only in the individuals raised in the harsher environment*" (emphasis mine).

This confirms that a benign environment delays ageing and reproduction, along the r-K selection model. The more safe and benign (but still positively challenging) the environment is, the more likely it is that ageing need not take place, as resources are no longer preferentially diverted to the germ line - there is no need, as the organism has reduced chances of early death. This and many other examples show that mechanisms for soma-to-germ line reallocation of resources exist in nature, are likely to be widespread (in insects, higher animals *etc.*), and can be made to operate by manipulating the environment to make it less dangerous albeit still mildly challenging.

To put it simply, these findings suggest once more that, on the whole, <u>when germ cells are healthy, somatic repair decreases, and when germ cells are stressed,</u>

somatic repair improves (Fig. **2**). This process depends on clues and factors from the environment, which reflect how likely it is that the organism will survive in that specific environment. Dangerous environments shift the emphasis of resource allocation to the germ line, whereas safe environments reverse this trend to allow for better somatic repair, and thus increased longevity. The process is made possible by the existence of cross-talk pathways and mechanisms between the two parties involved, a process reminiscent of the importance of having dialogue channels open in the political realm when there is the possibility of conflict [32].

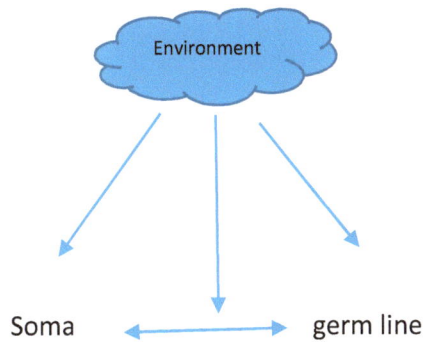

Fig. (2). A simplified diagram showing the impact of environmental challenges. In this diagram there is cross-talk between soma and germ line which, at present favours the germ line with increased repair resources but can, conceivably, be reversed.

The 'Indispensable Soma Hypothesis'

As mentioned above, evidence is gathering suggesting that that the disposable soma theory of ageing is not unidirectional and the soma may not, after all, be always 'disposable' [10]. Under certain conditions of evolutionary necessities there could be increased somatic maintenance at the expense of germ cells. This, translated into clinical terms, may mean that the soma (as defined in Box 1) experiences continual repair and lives without any significant age-related chronic degeneration. I have suggested [33] that some of these evolutionary pressures could be dependent upon how well one makes themselves 'indispensable' within the global techno-cultural environment, through intense (but hormetic) sharing of information, which has an impact on somatic neurons which, then, diminish the evolutionary importance of germ line cells (Fig. **3**).

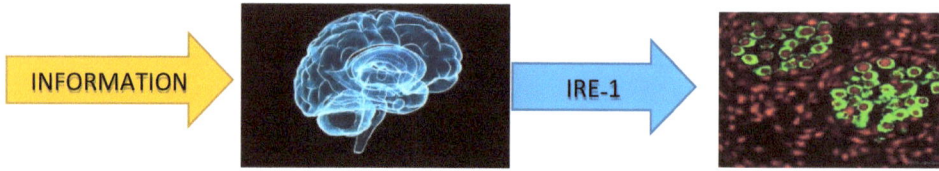

Fig. (3). Challenging information results in damaged germ line cells. Environmental high-quality information which creates a tendency for action, up-regulates human neurons, which generate IRE-1 (among many other stress response factors) which results in cell line cell apoptosis, an event which subsequently releases signals to upgrade the repair of somatic cells.

Ermolaeva *et al.* [34] present research which seems to support this view. Their report raises the possibility that DNA damage in germ cells may protect somatic cells. They suggested that DNA damage in **germ cells** can up-regulate stress resistance pathways in **somatic cells** and improve stress response to heat or oxidation. This is profoundly important because it shows that, in principle, when germ cells are damaged, they initiate a cascade of events and produce agents which can then protect somatic cells against systemic stress. It can be deducted that this can be a possible mechanism for the phase transition discussed in other chapters, because it provides the theoretical basis for the altered biological mechanisms. One somatic stress protection mechanism originating from germ cells is mediated through the MAP (mitogen activated protein)-kinase homologue MPK-1 in germ cells. This triggers other agents which are hitherto poorly described, to up-regulate elements of the Ubiquitin-Proteasome system in somatic cells. As a result, there is increase resistance to stress in somatic tissues. This mechanism may reflect an innate tendency to reverse the trade-offs between germ cell and somatic cell repair: when the germ cells are compromised, there is delay in offspring production matched by an increased repair of somatic cells [34].

Another possible biological mechanism involved in immortalisation has been described by De Vaux *et al.* [35]. They confirm that the Mi2 protein (a nucleosome-remodelling protein) causes repression of transcription and is involved in the repair of germ line DNA. They also suggest that a Mi2 homolog in *C.elegans* called LET-418/Mi2 is one main factor which drives development and reproduction in early life and then causes dysfunction in later life. Inactivation of LET-418/Mi2 can result in increased resistance to stress and enhanced longevity:

therefore it can regulate lifespan. The authors suggest that this agent is involved in germ cell apoptosis. This provides a biological basis for a possible mechanism which could be potentially subjected to modulation, in order to produce life-span prolongation. It also supports the general concept that germ line dysfunction results in somatic lifespan extension as a counter-effect.

In summary then, and in simple terms:

1. Healthy germ cells tend to divert repair resources away from somatic tissues, in order to improve their own survival.
2. Challenged or 'positively-stressed' somatic neurons initiate injury and apoptosis in germ cells.
3. Damaged germ cells may produce agents which protect somatic tissues.

This makes perfect evolutionary sense: If germ line cells are damaged (and thus the survival of the species is at risk), then an effective alternative would be to achieve survival through a longer and more efficient somatic repair. From the point of view of evolution, if the species cannot survive, then at least the individual bodies should. In fact, this is the original plan embedded in natural laws. Research in this area is slowly unravelling the different mechanisms involved. Overall, it seems plausible, though speculative, that by acting in a way that increases one's value in the global landscape (and this may be achieved *via* continual, hormetic exposure to information and challenges), one is subjected to biological and evolutionary mechanisms which increase one's survival. This is the **'indispensable soma hypothesis'**: a soma which makes itself indispensable by acting positively within the general process of successful adaptation, may experience a reversal of the effects that govern the disposable soma theory, and thus live longer. In reality, the most likely agent able to achieve such a scenario is the human neuron and its supporting tissues, *i.e.* entire non-reproducing techno-culturally aligned humans [36].

The Role of Transposable Elements in Soma to Germ Line Communication

Let me examine some further details regarding the possible operation of these mechanisms. One fruitful and promising area of research in this respect, is the study of transposable elements. Transposable elements, or transposons, are DNA sequences which may migrate and take new positions within the genome. This may result in new functions or new malfunctions, for example due to the creation of mutations or due to reversal of the effects of existing mutations (Fig. **4**).

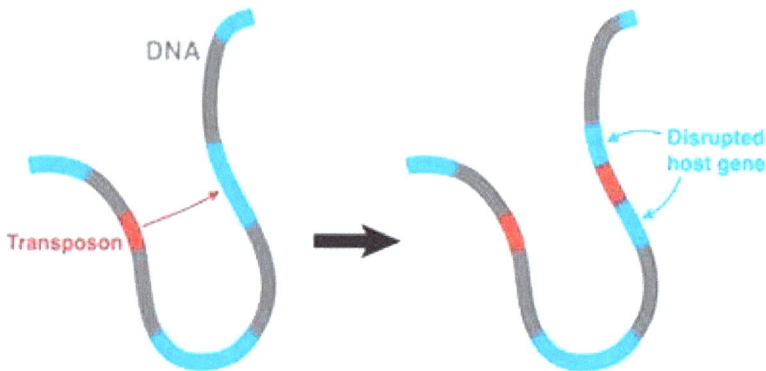

Fig. (4). The transposition of a genetic element within the genome. Transposons can replicate and then insert themselves randomly throughout the genome. This can cause mutations which can be negative, neutral or positive, depending of the function being studied (Illustration by Phillip Dumesic, UCSF. https://www.ucsf.edu/news/2013/02/13535/gene-invaders-are-stymied-cells-genome-defense.

Regarding one possible epigenetically-based mechanism of immortalisation, Smelick and Ahmed [7] quote:

> *"Maintenance of epigenetic silencing of repetitive elements such as transposons is another critical feature of the programming process that is essential for genome stability. Carefully co-ordinated epigenetic silencing and reprogramming could, therefore, constitute a chromatin maintenance mechanism that contributes to germline immortality".*

This implies that is such a mechanism exists it could, in principle, be applied on somatic cells if it is required by evolutionary necessities. The role of transposons in germ line immortality has been examined by Fontana who introduced the concept of Epigenetic Tracking as mentioned above [15]. This is an evolutionary method of generating complexity. He suggests that transposition in somatic cells drive cellular differentiation and development, and transposition in germ cells facilitates evolution. This implies that regulation of transposons both in soma and germ line may result in effective damage repair and improvement of function. I posit that positive regulation in somatic transposons may be achieved through the mechanisms that regulate the assimilation of novel information by the cell. We know that transposon expression is responsive to challenging ambient stimulants, and that cells undergoing hormetic stress may trigger transposon activity. Transposons act both locally and globally through microRNAs, and in this way behave as regulators of the stress response [37]. It is also known that genetic manipulation using transposons, such as the sleeping beauty (SB) transposon [38] is a useful approach in the study of mutagenesis and genetic behaviour, both in the soma and in the germ line. It is possible, though speculative, that fractions of ancient genomes that accumulated specific mutations, (but that were able to repair somatic damage) exhibited the phenomenon of 'stochastic loss' whereby active transposons may have completely disappeared from the genome. However, this loss has no significance in the germ line because horizontal transfer is capable of rescuing any active transposons which start exhibiting stochastic loss. As a result, the genome of somatic cells became progressively less stable, whereas the genome of germ line cells remained active and fully functioning. This is a useful concept which indicates, once again, a clear genomic dichotomy between soma and germ line, a dichotomy which lends itself to manipulation and possible reversal. As the SB transposon forms part of the Tc1/mariner superfamily [39], members of which are very widespread in nature, there is no reason to believe that its actions are rare or bear no significance in the case we are examining (*i.e.* in human somatic maintenance). Therefore, transposons including the well-studied SB transposon may be a fruitful area of research which can elucidate senescence-reversing mechanisms. According to Ivics *et al.* [38]:

> *Protecting the genome from transposable element (TE) mobilization is critical for germline development. In Drosophila, Piwi proteins and their bound small RNAs (piRNAs) provide a potent defense against TE activity. TE targeting piRNAs are processed from TE-dense heterochromatic loci termed piRNA clusters. Although piRNA biogenesis from cluster precursors is beginning to be understood,*

little is known about piRNA cluster transcriptional regulation. Here, we show that deposition of histone 3 lysine 9 by the methyltransferase dSETDB1 (egg) is required for piRNA cluster transcription. In the absence of dSETDB1, cluster precursor transcription collapses in germline and somatic gonadal cells and TEs are activated, resulting in germline loss and a block in germline stem cell differentiation. We propose that heterochromatin protects the germline by activating the piRNA pathway.

This highlights some further details regarding the role of transposons and microRNAs in germ line- somatic cross-talk. Suppressing transposon activity hinders germ line development and this may then incite germ cells to produce somatic-protecting factors as discussed above. Karnaukhov and Karnaukhova [40] have examined two processes leading to germ line immortality, through correction of genomic errors. They quote:

In 2009 we proposed a solution for the problem of "escaping germ line from aging" within the framework of a wide range of hypotheses based on the aging notion as a process of accumulation of genomic errors.... It has been found that the mechanism providing "correction" of genomic errors in germ line cells involves two elements which are well known and common practically for all eukaryotic organisms: Gene recombination (crossing over) ...(and gamete selection). It is well known that as a result of gene recombination (crossing over) in cells of germ line the haploid daughter cells-precursors are replicated. These haploid daughter cells (gametes) contain 1/2 of the genetic information of next generation organism. Need to note here, that the density of gamete's genomic errors differ from the density of genomic errors in germ line cells of maternal organism. Indeed, the stochastic nature of the location of genomic errors and the stochastic nature of gene recombination (crossover) leads to the random distribution of errors between gamete's genomes. And it is important that the density of the genomic errors in some of them is lower than in the cells of germ line of maternal organism.... The given mechanism of "correction" of the genomic errors is the process of selection of precursor cells (gametes) with minimal error density. This becomes possible of an excessive number of gametes. Moreover, the haploidy of such cells raises the efficiency selection inasmuch as it elevates the specific contribution of every error into the overall reduction of gamete functionality. Although each element of this mechanism (gene recombination + gamete selection) was in itself well known and studied, the conclusion that the reduction of the density of genomic errors comes out from their joint action, was a priority of our previous works. This solution of the paradox of the "not aging germ line" rehabilitates a wide range of aging hypotheses united by the idea of degradation of the genetic information by aging.

Their comment is a reminder of some mechanisms of genomic error correction in the germ line. In this case, the correction of errors is due to a combined action of gene recombination and gamete selection, both of which could also be made to operate in somatic cells. Of course, the element of speculation here is considerable, although not extreme. We know (or, at least, we can infer) that such mechanisms may be subjected to exaptation which can produce functions that were not present originally [for example 41]. Exaptation (the shift in the function of a trait during evolution, when a trait is co-opted for a different function than the original) may underlie some mechanisms which are involved in homoeostasis but it may also operate within the scenario presented in this book, which is the re-allocation of resources from germ line to soma. Those mechanisms originally developed in order to facilitate resource flow from soma to germ line (such as gene recombination and gamete selection), may (under certain circumstances, such as those discussed above) be exapted and operate in the opposite way: shift the balance of repair from germ line to soma [42]. This may not be a very long process, as it operates through epigenetic mechanisms some of which involve fast microRNA regulation [43].

CONCLUSION

This discussion builds on the previous analysis of the effects of hormesis and environmental enrichment, as it concentrates specifically upon the genetic effects of such stimulation. We may now be able to better appreciate that:

a. there is a bidirectional cross-talk between soma and germ line, and
b. factors and mechanisms which mediate (or can modify) the dynamics of this cross-talk are being increasingly elucidated.

Environmental factors may directly or indirectly influence epigenetic networks and ultimately influence chromatin and histone modifications resulting in altered protein expression. This research helps us conceive a facilitating mechanism where environmental information (*via* appropriate challenges) may affect germ line function, and provides a medium whereby we could potentially intervene and influence this function (ultimately aiming to redirect repair resources from germ line to soma). I reiterate here that the term 'somatic cell' referred to in this discussion is more likely to be primarily a neuron, and secondarily all cells that support the neuron's survival (*i.e.* essentially, a human organism surrounding a human brain). Although the mechanisms involved in this hypothesis need further elucidation, and the speculative inferences need further grounding, it may be possible to commence suggesting biologically-founded mechanisms which may underpin our earlier speculations: that the environment *via* epigenetic factors such as microRNA and transposons, can influence (cybernetically, 'nudge') the balance between somatic *vs* germ cell immortalisation, resulting in improved somatic repair, a reduction of age-related degeneration, and thus increased human lifespan.

REFERENCES

[1]   Lumey LH, Stein AD, Kahn HS, *et al.* Cohort profile: the Dutch Hunger Winter families study. Int J Epidemiol 2007; 36(6): 1196-204.
       [http://dx.doi.org/10.1093/ije/dym126] [PMID: 17591638]

[2]   Fredriksson Å, Johansson Krogh E, Hernebring M, *et al.* Effects of aging and reproduction on protein quality control in soma and gametes of *Drosophila melanogaster.* Aging Cell 2012; 11(4): 634-43.
       [http://dx.doi.org/10.1111/j.1474-9726.2012.00823.x] [PMID: 22507075]

[3]   Drenos F, Kirkwood TB. Modelling the disposable soma theory of ageing. Mech Ageing Dev 2005; 126(1): 99-103.
       [http://dx.doi.org/10.1016/j.mad.2004.09.026] [PMID: 15610767]

[4]   Gracida X, Eckmann CR. Fertility and germline stem cell maintenance under different diets requires nhr-114/HNF4 in *C. elegans*. Curr Biol 2013; 23(7): 607-13.
[http://dx.doi.org/10.1016/j.cub.2013.02.034] [PMID: 23499532]

[5]   Heininger K. Aging is a deprivation syndrome driven by a germ-soma conflict. Ageing Res Rev 2002; 1(3): 481-536.
[http://dx.doi.org/10.1016/S1568-1637(02)00015-6] [PMID: 12067599]

[6]   Chastain E, Antia R, Bergstrom CT. Defensive complexity and the phylogenetic conservation of immune control 2012.

[7]   Smelick C, Ahmed S. Achieving immortality in the *C. elegans* germline. Ageing Res Rev 2005; 4(1): 67-82.
[http://dx.doi.org/10.1016/j.arr.2004.09.002] [PMID: 15619471]

[8]   Partridge L, Gems D, Withers DJ. Sex and death: what is the connection? Cell 2005; 120(4): 461-72.
[http://dx.doi.org/10.1016/j.cell.2005.01.026] [PMID: 15734679]

[9]   Charlesworth B. Fisher, Medawar, Hamilton and the evolution of aging. Genetics 2000; 156(3): 927-31.
[PMID: 11063673]

[10]   Douglas PM, Dillin A. The disposable soma theory of aging in reverse. Cell Res 2014; 24(1): 7-8.
[http://dx.doi.org/10.1038/cr.2013.148] [PMID: 24189044]

[11]   Curran SP, Wu X, Riedel CG, Ruvkun G. A soma-to-germline transformation in long-lived *Caenorhabditis elegans* mutants. Nature 2009; 459(7250): 1079-84.
[http://dx.doi.org/10.1038/nature08106] [PMID: 19506556]

[12]   Petruseva IO, Evdokimov AN, Lavrik OI. Molecular mechanism of global genome nucleotide excision repair. Acta Naturae 2014; 6(1): 23-34.
[PMID: 24772324]

[13]   Murphey P, McLean DJ, McMahan CA, Walter CA, McCarrey JR. Enhanced genetic integrity in mouse germ cells. Biol Reprod 2013; 88(1): 6.
[http://dx.doi.org/10.1095/biolreprod.112.103481] [PMID: 23153565]

[14]   Kirkwood TB. Immortality of the germ-line *versus* disposability of the soma. Basic Life Sci 1987; 42: 209-18.
[PMID: 3435387]

[15]   Fontana A. A hypothesis on the role of transposons. Biosystems 2010; 101(3): 187-93.
[http://dx.doi.org/10.1016/j.biosystems.2010.07.002] [PMID: 20655980]

[16]   Levi-Ferber M, Salzberg Y, Safra M, Haviv-Chesner A, Bülow HE, Henis-Korenblit S. It's all in your mind: determining germ cell fate by neuronal IRE-1 in *C. elegans*. PLoS Genet 2014; 10(10): e1004747.
[http://dx.doi.org/10.1371/journal.pgen.1004747] [PMID: 25340700]

[17]   Prause MC, Berchtold LA, Urizar AI, *et al.* TRAF2 mediates JNK and STAT3 activation in response to IL-1β and IFNγ and facilitates apoptotic death of insulin-producing β-cells. Mol Cell Endocrinol 2015; S0303-7207(15): 30146-5.

[18]   Wu Q, Cao X, Yan D, Wang D, Aballay A. Genetic screen reveals link between maternal-effect sterile gene mes-1 and P. aeruginosa-Induced neurodegeneration in *C. elegans*. J Biol Chem 2015.

[19]   Sharma A. Novel transcriptome data analysis implicates circulating microRNAs in epigenetic inheritance in mammals. Gene 2014; 538(2): 366-72.
[http://dx.doi.org/10.1016/j.gene.2014.01.051] [PMID: 24487054]

[20]   Smendziuk CM, Messenberg A, Vogl AW, Tanentzapf G. Bi-directional gap junction-mediated soma-germline communication is essential for spermatogenesis. Development 2015; 142(15): 2598-609.
[http://dx.doi.org/10.1242/dev.123448] [PMID: 26116660]

[21]   Sharma A. Transgenerational epigenetic inheritance: resolving uncertainty and evolving biology. Biomol Concepts 2015; 6(2): 87-103.
[http://dx.doi.org/10.1515/bmc-2015-0005] [PMID: 25898397]

[22]   Torres-Padilla ME, Ciosk R. A germline-centric view of cell fate commitment, reprogramming and immortality. Development 2013; 140(3): 487-91.
[http://dx.doi.org/10.1242/dev.087577] [PMID: 23293280]

[23]   Sharma A. Bioinformatic analysis revealing association of exosomal mRNAs and proteins in epigenetic inheritance. J Theor Biol 2014; 357: 143-9.
[http://dx.doi.org/10.1016/j.jtbi.2014.05.019] [PMID: 24859414]

[24]   Rangan P, Malone CD, Navarro C, *et al.* piRNA production requires heterochromatin formation in *Drosophila.* Curr Biol 2011; 21(16): 1373-9.
[http://dx.doi.org/10.1016/j.cub.2011.06.057] [PMID: 21820311]

[25]   Tan JY, Sirey T, Honti F, *et al.* Extensive microRNA-mediated crosstalk between lncRNAs and mRNAs in mouse embryonic stem cells. Genome Res 2015; 25(5): 655-66.
[http://dx.doi.org/10.1101/gr.181974.114] [PMID: 25792609]

[26]   Kenyon C. A pathway that links reproductive status to lifespan in *Caenorhabditis elegans.* Ann N Y Acad Sci 2010; 1204: 156-62.
[http://dx.doi.org/10.1111/j.1749-6632.2010.05640.x] [PMID: 20738286]

[27]   Khodakarami A, Saez I, Mels J, Vilchez D. Mediation of organismal aging and somatic proteostasis by the germline. Front Mol Biosci 2015; 2: 3.
[http://dx.doi.org/10.3389/fmolb.2015.00003] [PMID: 25988171]

[28]   Shemesh N, Shai N, Ben-Zvi A. Germline stem cell arrest inhibits the collapse of somatic proteostasis early in *Caenorhabditis elegans* adulthood. Aging Cell 2013; 12(5): 814-22.
[http://dx.doi.org/10.1111/acel.12110] [PMID: 23734734]

[29]   Bazley FA, Liu CF, Yuan X, *et al.* Direct reprogramming of human primordial germ cells into induced pluripotent stem cells: efficient generation of genetically engineered germ cells. Stem Cells Dev 2015; 24(22): 2634-48.
[http://dx.doi.org/10.1089/scd.2015.0100] [PMID: 26154167]

[30]   Nagamatsu G, Kosaka T, Saito S, *et al.* Induction of pluripotent stem cells from primordial germ cells by single reprogramming factors. Stem Cells 2013; 31(3): 479-87.
[http://dx.doi.org/10.1002/stem.1303] [PMID: 23255173]

[31]   Kim SY, Metcalfe NB, Velando A. A benign juvenile environment reduces the strength of antagonistic pleiotropy and genetic variation in the rate of senescence. J Ecol 2015. [http://dx.doi.org/10.1111/1365-2656.12468] [PMID: 26559495]

[32]   Navaro-Yashin Y. Pacifist devices: the human/technology interface in the field of 'conflict resolution'. Cambridge J Anthropol 2008; 28(30): 91-112.

[33]   Kyriazis M. Technological integration and hyperconnectivity: Tools for promoting extreme human lifespans. Complexity 2015; 20(6): 15-24. [http://dx.doi.org/10.1002/cplx.21626]

[34]   Ermolaeva MA, Segref A, Dakhovnik A, *et al.* DNA damage in germ cells induces an innate immune response that triggers systemic stress resistance. Nature 2013; 501(7467): 416-20. [http://dx.doi.org/10.1038/nature12452] [PMID: 23975097]

[35]   De Vaux V, Pfefferli C, Passannante M, *et al.* The *Caenorhabditis elegans* LET-418/Mi2 plays a conserved role in lifespan regulation. Aging Cell 2013; 12(6): 1012-20. [http://dx.doi.org/10.1111/acel.12129] [PMID: 23815345]

[36]   Last C. Human evolution, life history theory, and the end of biological reproduction. Curr Aging Sci 2014; 7(1): 17-24. [http://dx.doi.org/10.2174/1874609807666140521101610] [PMID: 24852016]

[37]   Wheeler BS. Small RNAs, big impact: small RNA pathways in transposon control and their effect on the host stress response. Chromosome Res 2013; 21(6-7): 587-600. [http://dx.doi.org/10.1007/s10577-013-9394-4] [PMID: 24254230]

[38]   Ivics Z, Kaufman CD, Zayed H, Miskey C, Walisko O, Izsvák Z. The sleeping beauty transposable element: evolution, regulation and genetic applications. Curr Issues Mol Biol 2004; 6(1): 43-55. [PMID: 14632258]

[39]   Ivics Z, Izsvák Z. Sleeping beauty transposition. Microbiol Spectr 2015; 3(2) [http://dx.doi.org/10.1128/microbiolspec.MDNA3-0042-2014]

[40]   Karnaukhov AV, Karnaukhova EV. [Informational hypothesis of aging: how does the germ line "avoid" the aging?]. Biofizika 2009; 54(4): 726-32. [PMID: 19795796]

[41]   Papamichos SI, Margaritis D, Kotsianidis I. Adaptive evolution coupled with retrotransposon exaptation allowed for the generation of a human-protein-specific coding gene that promotes cancer cell proliferation and metastasis in both haematological malignancies and solid tumours: the extraordinary caseof MYEOV gene. Scientifica (Cairo) 2015; 2015: 984706.

[42]   Singh DP, Saudemont B, Guglielmi G, *et al.* Genome-defence small RNAs exapted for epigenetic mating-type inheritance. Nature 2014; 509(7501): 447-52. [http://dx.doi.org/10.1038/nature13318] [PMID: 24805235]

[43]   Kannan S, Chernikova D, Rogozin IB, *et al.* Transposable element insertions in long Intergenic non-coding RNA genes. Front Bioeng Biotechnol 2015; 3: 71. [http://dx.doi.org/10.3389/fbioe.2015.00071] [PMID: 26106594]

CHAPTER 6

Another Dimension: 'Zooming Out'

Abstract: The quest to find effective therapies aimed at chronic age-related degeneration has deep and wide ramifications. It is not sufficient to examine physical or pharmacological interventions which may have an impact on the process. Instead, we also need to consider more profound evolutionary principles which underpin the process of ageing. One example is the principle of degeneracy which may be useful in explaining how we may attain similar functions by using different structures. Another example is hysteresis, which examines dependence on already established conditions - a crucial obstacle we need to overcome in the quest to diminish chronic degeneration. This chapter will be an exploration of certain evolutionary and philosophical principles, such as a contemplation of the meaning of life, which complement the biological and medical ones. Concepts relating to resilience, complexity and self-organisation, as well as a discussion of certain cybernetic principles (such as path-dependency and nudging), all taken together will provide a suitable and realistic framework for achieving our aim: to manipulate nature in a way that diminishes the impact of age-related degeneration, and reduces mortality as a function of age.

Keywords: Boundary, Complexity, Degeneracy, Health, Homoeodynamic space, Hysteresis, Life, Lifestyle, Nudging, Path dependency, Resilience, Resilient interface, Stigmergy.

INTRODUCTION

In order to place the general discussion of this book into a broader framework, it is necessary to reflect on some concepts which hitherto have not been considered with due vigour by gerontologists (Fig. **1**).

These concepts have nevertheless, direct and significant relevance upon the ageing process. The question whether there is an ultimate 'aim' or 'purpose' in evolution has tormented the greatest minds ever lived. In order to simplify the matter, and also in order to fashion the concept into something relevant to this book, I posit that:

1. Nature has a general tendency to 'life'. Evolution is biased towards increasing fitness in order to ensure survival.
2. But, in humans this tendency has reached another level. Human creativity has evolved into something that it is clearly over and above what is needed to ensure mere survival.

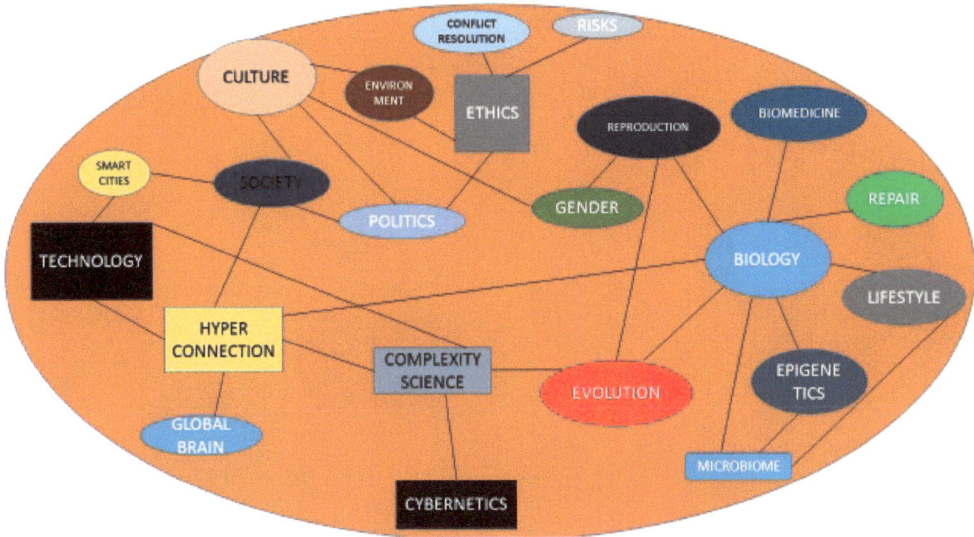

Fig. (1). A wider picture of ageing. In order to devise methodologies which may help us eliminate age-related degeneration, we need to see a bigger picture, and consider ageing as a mechanism which is being influenced by many diverse concepts. The disciplines of biology or medicine are not enough. We need to study societal and environmental aspects, complexity principles, human evolution, ethics and politics and a generally much wider approach.

This creative impulse has been identified by some philosophers as the effort to 'transform darkness into light' [1]. It is in other words, a quest, to continually reach a stage which is better compared to the previous one. Organic agents and humans in particular, are merely instruments of nature, existing only in order to enhance this fundamentally-driven process of continual evolution. These agents are then discarded, exchanged for new ones (new individuals, new societies, new cultures) in perpetuity. Each participating organism contributes a small part in the entire process, and then it perishes. There is nothing in this scenario to suggest that the duration of this presence must be short. In fact it can be as short or as long as necessary in order to facilitate and enhance the adaptive capabilities of the entire mechanism of continual evolution.

The view put forward in this book is not merely to just improve our health in order to increase our survival. No. It is to improve our health which will lead to our prolonged survival, in order to achieve a superior stage of evolution, and so fulfil the innate creative impulse of humanity (Figs. **2** and **3**).

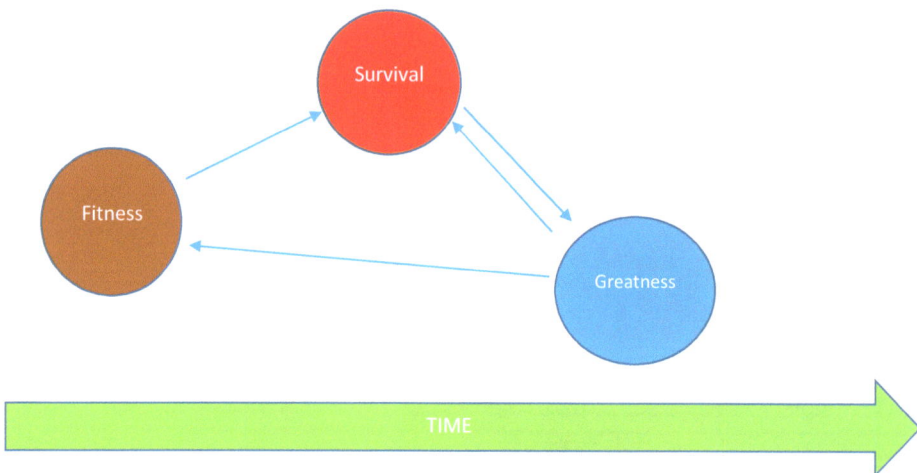

Fig. (2). Human destiny over time. In this example, 'fitness' is literal *i.e.* 'being fit', appropriate and suitable for one's environment. 'Greatness' (intellectual creativity) is defined as: 'the continual quest to reach a stage which is better compared to the previous one'. Being fit increases the chances of long life, which improves the chances of achieving a higher stage of evolution. This, in turn, helps to increase fitness and further improves survival, in a continual manner. This process is fuelled by the continual input of meaningful, organised information (as in Fig. **3**).

Living systems are defined by their ability to contain high level of information [2]. Any far-from- equilibrium systems (such as living humans) have high entropy generation, and the way to maintain a stable entropic state is to achieve an extreme level of information content [3]. Frieden and Gatenby [3] quote:

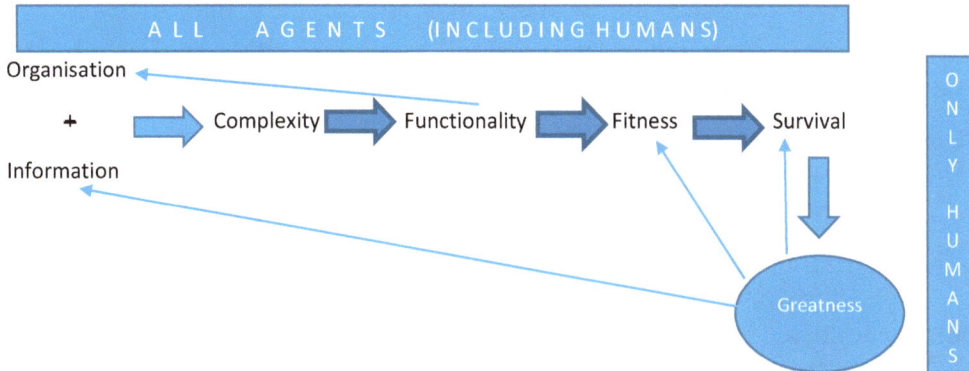

Fig. (3). How information impacts on human greatness. This is an extension of Fig. (1) discussed in chapter 4: With specific reference to humans, we can now add an additional layer of complexity, over and above mere survival. This is the innate tendency of humans to continually reach for something superior. As we reach higher levels of creativity, we also improve the flow of information in the system, as well as optimise our fitness and survival, reciprocally.

"We have previously demonstrated that information in a biological context can be viewed as the capacity to facilitate work. Specifically, it directs and catalyzes the conversion of energy and substrate from the environment into specific macromolecules that, in turn, maintain the orderly structure of the cell...The stability of a living system requires its information content to be maintained at an extremum...Information converts environmental energy to intracellular order... Studies aiming to see how normal cells change into cancerous, suggest that the process represents an: ... *information phase transition from a maximum to a minimum value,* probably *through a number of unstable intermediates. Our model indicates this transition could be initiated by* loss of energy *through intra- or extra-cellular factors"...* (**emphasis mine**).

It is important to appreciate that evolution favours increased functional complexity, as both antagonistic interactions and natural selection have been shown to tend towards complexity and evolvability [4, 5]. If we accept that there is a natural propensity to increase functional complexity (which improves fitness and survival) then it becomes more likely that any process that increases complexity (*via* increased challenges as discussed throughout this book), can be preferentially selected by evolutionary mechanisms. Survival depends on the ability of agents (entities which act on their environment) to capture fitness-relevant features from their constantly changing habitat. In the case of modern humans, our ability to adapt quickly to a rapidly changing, information-laden environment is crucial in this respect.

In this book we discussed how information form the environment is translated into biological actions. This information helps enrich the process in the wider sense (Fig. **3**), and it results in an improvement, an adaptation, while maintaining homoeostasis. At this point it would be helpful to discuss some general principles relating to the notion of resilience and how our internal world is interfacing with the environment: the concept of the resilient interface.

The Resilient Interface

Between ourselves and the environment there is an abstract interface which helps to minimize friction, conflict and interference. This may also be extended to include energy barriers which work in order to reduce loss of energy to the environment (*i.e.* to minimize entropy). The concept of the 'Resilient Interface' (RI) has been discussed in some detail by Shima Beigi, a resilience and sustainability scientist at the University of Bristol [6]. RI is the interface between an agent and its living space. The interface contains information about self-organisation and learning patterns of the agents as these interact with their environment. It characterises the resilience of an agent, and it helps us, in the case of ageing, to explore how the environment can influence our biology. Information embedded in the RI defines how an agent may react to any external challenging stimuli. In humans, RI can be conceptualised as a complex environment, including biological, mechanical, cultural and social elements, which are autonomous entities themselves, within an autonomous entity (the human body).

Examples of interfaces are:

- The skin tissue (including the mucosa) and the skin microbiota, forming an interface between the external and the internal environment [7]. This acts not only as a physical barrier but also as an immunological one, with close relationships and continual cross-talk between environmental challenges, the skin microbiota and the immune cells inside the organism [8].

- The extracellular matrix, which is increasingly being recognized as a dynamic system that interferes both with the function of cells and the rest of the body [9].
- The Blood Brain Barrier, the dynamic physico-chemical interface between brain tissues and the blood [10].
- The gut microbiota which is the interface between the gut (nutritional) environment and the internal milieu of our body [11].
- The metabolome, small organic molecules which form an abstract interface between our metabolic processes and the external environment [12].
- The immune system, the dynamical barrier between external infective agents and allergens, and the internal milieu (for example [13]).
- The vagina, an erotic/physicochemical barrier with its own microbiome, a RI both in the biological and socio-cultural sense [14].
- The placenta, a temporary but important interface, which regulates development.

The flow of data between the internal and the external world is facilitated, modified and modulated by the RI. The definition of 'Resilience' is: *the ability of a system or any agent to learn, grow or develop (a positive re-introduction of order) following any perturbations that disturb that system or agent.* The resilient interface should mediate between agent and its living space in order to minimize friction, reduce conflict, facilitate co-operation, regulate noise, and decrease interference.

At this point is worth revisiting some concepts discussed by Chatterjee in chapter 7. He considers a human organism as a Complex Adaptive System (CAS), which is separated from the surrounding space by a virtual **boundary.** This is implied by the Laws of Thermodynamics, as transfers of entropy and energy cross this boundary in a reciprocal fashion. He highlights that: "A *system boundary gives a structure to the system, and also acts as a separating layer between the surrounding environment and the system's internal machinery*". The interactions that happen inside the organism result in generation of energy or entropy which then dissipates across the boundaries. Thus, there is a border, a limit between the system (the collection of interacting components) and the external environment.

This interface allows for the presence of **gradients** which exist in order to limit dissipation of energy by the system organism. It is also important to revisit the concept of 'closure' examined Chapter 7. 'Closure' is a property of organised systems, when a causal chain of actions eventually closes onto itself, so that a particular function is performed. *"Functional closure is defined as a closed cycle of processes that do not form a physical mediating layer, such as catalytic reactions in an autocatalytic set. In contrast to this, structural closure demands that the interactions create a physically closed topology."* These concepts are extremely relevant in a discussion about ageing, as they explain how a system may malfunction due to a failure of its various interfaces, its boundaries, its gradients, its buffering actions, and its ability to close an already initiated interaction. These failures happen in ageing and are due to an underlying general increase in entropy. The relevance here is that we can now begin to discuss ways to prevent or reverse such failures of boundaries or interfaces, which destroy homoeostasis, damage our buffering abilities, and disrupt our homeodynamic space.

Beyond Homoeostasis: Degeneracy, Robustness and Resilience

Human metabolic systems conceal multiple pre-adaptive traits which may evolve into innovative and adaptive functions [15]. Thus, the potential to evolve and adapt is present in all of us, and there are ways by which it can be made to work to our advantage. One way this may be operating is through the concept of degeneracy.

A. Degeneracy

The capacity of alternate structures to achieve similar functional results in one context, or alternative functional outcomes in a different context, is called degeneracy. Degenerate elements have a 'structure-to-function' ratio of one-to-many, *i.e.* they are pluripotent. In degenerate systems there is separation of function. Elements suited to one function may - within a particular environment-exhibit a divergence of function and adopt a different function [16]. These elements are 'exapted' to a new function without interfering with their original system. Therefore, if the environment **requires** a new function (such as resource investment in cognition) then our degenerate systems will provide this function [17]. One vehicle for this biological change is the epigenome which can help in the evolution of new phenotypes from the same original genes. In this respect, degeneracy prevents a cascade of failures, and it helps improve plasticity and adaptability of a system [18].

Tononi *et al.* [19] have suggested that any system with high degree of degeneracy is generally able to integrate information at a high level. This supports the view that complex organisms which are able to deploy degenerate mechanisms (such as humans), evolve better within a complex environment. These organisms are better able to respond to novelty, new challenges or stressful events, exactly because several (perhaps unrelated) structures and elements will converge (degenerate) their function in order to respond to the novel stimulus.

The relevance of degeneracy in this discussion is that our homoeostasis may shift its priorities into a new phenotype **if** our environment requires it: due to the fact that our modern techno-culture requires high levels of cognition (a challenging, power-law dependent level of information-processing), and there is an increased need to adapt and evolve quickly, it is foreseeable that this function will be made available through our degenerate systems. I am speculating that this will not involve any new and additional resource or energy problems as the resources will merely need to be diverted, from one system (germ) to another (soma), where they are required by the underlying forces of evolution.

B. Robustness and homoeostasis

Robustness is a property that allows a system to maintain its functions against internal and external perturbations [20]. Robustness does not refer to system states at any specific moment. Instead it refers to the ability to maintain a given function. Thus robustness does not refer to a sound homoeostasis, but it is the ability to maintain homeostasis. Kitano [20] describes homeostasis as follows:

"The coordinated physiological processes which maintain most of the steady states in the organism are so complex and so peculiar to living beings—involving, as they may, the brain and nerves, the heart, lungs, kidneys, and spleen, all working cooperatively—that I have suggested a special designation for these states, homeostasis. The word does not imply something set and immobile, a stagnation. It means a condition—a condition which may vary, but which is relatively constant'..."

That is to say, homeostasis is a "property that maintains the state of the system rather than its functions". Thus, robustness (of function) contributes to overall homoeostasis. Rattan [21] clarifies this better by using the concept of the "homeodynamic space" which is the space in which the system can move so that it can adapt to any disturbances. This space progressively shrinks with age, resulting in increased fragility (and loss of function) so this must be reversed or at least its rate of degeneration to be slowed down. Kitano [20] also provides a mathematical foundation of robustness of biological systems, and aims to study underlying principles associated with it. He quotes further:

> "Living organisms are designed through evolution and perturbed under environmental constraints. Each instance of design is an actual life form that exists in the past, present, and future. Viable design is only possible within the constraints of fundamental principles and structural principles. Fundamental principles include basic laws such as quantum theory, Maxwell's equations, basic chemistry, and physics that apply to almost everything universally. Structural principles govern properties of systems and have a specific architecture such as control theory, communication theory, and various principles applied to specific configurations of components that are generally architecture-specific and context-dependent. For systems biology to be truly successful, not only studies on specific instances of life, but also studies on principles governing the entire design space are required."

These concepts are absolutely essential in the study of ageing, as they confirm, once more, that the processes involved in ageing depend on basic, fundamental laws and principles, which are not only biological but also embrace other disciplines such as physics and chemistry. We cannot hope to address ageing unless we also consider all of these, and other concepts including social, ethical, political and philosophical [22] (Fig. **1**).

Kitano, furthermore, says: *"Highly optimized tolerance (HOT) theory demonstrates that a system that is optimized for a <u>specific</u> perturbation inevitably entails extreme fragility for <u>unexpected</u> perturbations."* Therefore, it is essential to maintain tolerance to a wide range of perturbations (by developing this *via* continual exposure to novel challenges) [23]. This also highlights the importance of not over-challenging the system, but using the right amount of stimulation required in order to elicit a positive action. A robustness to perturbations leads to good homoeostasis. But homeostasis changes, depending on environmental constraints or conditions. Despite the name, homoeostasis is not static and inert, but it shifts to states which reflect an optimum reaction to the external conditions-the system adapts to any external variable. However, it is important to consider the issue of trade-offs. Any stimulation which may lead to a positive event is associated with a trade-off in another part of the system. For instance, a drug with a strong therapeutic effect may be associated with serious side effects. Trade-offs are inherent in any system and cannot be evaded, but their effect can be mitigated, by selecting trade-offs which are not entirely adverse to the function of the system. Any gain of robustness in one area results in increased fragility in another, but if this particular area can tolerate the fragility, then the severity of the trade-off is softened.

The problem here is that by trying to increase the robustness of a system, we are also increasing the energy and resources demand on that particular system. This makes the system more fragile (less robust) against **unpredicted** perturbations. This may be counteracted by optimising resource allocation, which may then result in both increased robustness and reduced fragility.

In this respect it could be argued that, using optimal resource allocation, we can achieve a high level of robustness in somatic tissues (which will then exhibit extended longevity), and **if** and when there are any unexpected or unplanned future perturbations which increase the somatic fragility (the trade-off between robustness and resources), **then** we can rely on artificial, externally-derived biomedical technology to counteract this fragility. Furthermore, the issue of performance is relevant. We can allow some trade-offs between performance and robustness (*i.e.* a reduced performance exchanged for improved robustness) provided that this reduction in performance is then supplemented by external technological means: we may use technology in order to supplement our reduced performance. This is a relevant example of how biology and technology merge in order to improve function. For example: Humans are now gaining cognitive robustness at the expense of losing physical robustness. There is a trade-off between cognition and physical strength. By becoming cognitively robust, we become physically 'fragile'. We are progressively losing our ability to run long distances, hunt and kill animals for food, fight in order to protect ourselves, and be able survive in the wild. However, this does not affect our function because our technology has provided supermarkets and supply chains, efficient transportation, and an increasingly secure social organisation, all of which fulfil the deficiencies in our physical function, and cancel out the effects of these particular trade-offs.

More on the Homoedynamic Space and the Phase Transition

Now, let me return to the concept of the homoedynamic space once again. Here we can develop further the notion of an 'attractor'. The topology of the homoedynamic space is characterised by the presence of attractors: points where systems tend to evolve towards. This is a useful notion because it helps us conceptualise what happens when we try to increase the information/energy of a system and how this increase can be facilitated by the proximity to the attractor. For instance, with regards to the germ line-to-soma transition of resources, consider the following schematic representation of energy requirements *vs* intervention (Fig. **4**).

Considering this concept form another point of view, it is known that a chronic or strong stressful event may cause a topological phase transition with radical rearrangements and remodelling of the structure and function of the network [24]. In the case of germ line cells, it may be that a chronic stress (one dependant on somatic cells constantly demanding resources for their own repairs) causes the germ line cell repair network to undergo a phase transition where fewer resources will be allocated to it and thus more will be diverted to somatic tissues. This shift from a resource-rich to a resource-poor environment may be crucial in inducing the transition from a functioning germ line cell to a non-functioning one. If the disturbance is sustained, then germ line cell repair networks may be down-regulated and disintegrate or at least attain lower function, so more emphasis will be placed on the somatic elements instead.

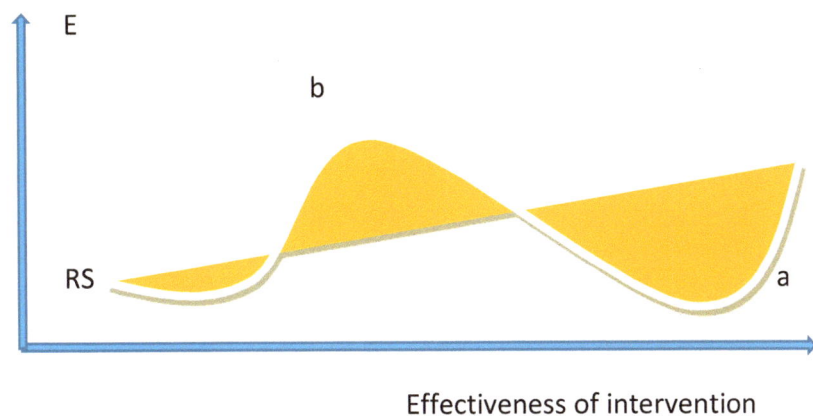

Effectiveness of intervention

Fig. (4). The dynamics of energy requirement in order to achieve function. Here, E= Energy required to overcome the barrier and cause the phase transition, RS= repair of somatic cells (as opposed to germ cell repair), b= the energy barrier, a= attractor. After reaching the peak of the barrier (b), the difficulty becomes less relevant as we move towards the attractor. Once we overcome the energy barrier, it becomes easier to achieve the primary aim (easier in energy terms). This suggests that it may not be impossible to transit from one state to another, as long as we devise an effective initial intervention, which will drive the system towards the attractor. Overcoming this energy barrier may be easier than generally believed, and this may be based on speculative quantum tunnelling effects.

To mention an example of proof-in-principle, Flatt *et al.* [25] have shown that changes in the honey bee environment have a direct effect upon the regulation of somatic maintenance. They suggest that the same genotype can express either of two alternative phenotypes, depending on the particular environment. Remaining with the honey bee example, Munch *et al.* [26] have shown that social cues can change a phenotype from a short-lived to a long-lived one. Social manipulation and changes in the environment may result in a slower rate of ageing and a slower accumulation of lipofuscin.

These examples support in principle the suggestion that suitable manipulation of the environment (*via* exposure to relevant digital information, for instance) can have epigenetic effects which divert resources away from the germ line and towards the soma, **if** this reversal is justified in evolutionary terms. In the case under consideration here, *i.e.* the reversal of soma-to germ line priorities in humans, I submit that the reversal is evolutionarily justified because the importance of a well-connected and well-functioning somatic neuron is greater compared to that of a germ cell, within a highly technological, cognition-dependant society. Calvanese *et al.* [27] suggest that the role of epigenetic modifications during ageing is defined by the need for these specific epigenetic changes to exist. I argue that in a highly information-rich environment it would be more relevant for somatic neurons (and as a consequence, other somatic cells that support the neuron) to be repaired and have priority over the repair of germ line elements. Having said that, we now encounter another theoretical obstacle to this transition: Hysteresis.

Hysteresis and Path-Dependency: an Obstacle in the Desired Phase Transition

The issue of the 'barrier' which needs to be overcome in order to achieve a reversal of resource allocations (from germ line to soma) can be considered under a different perspective. Here we are dealing with 'path-dependency' where the biological path of resources converging from the soma to the germ line has already been established by evolution. In order to reverse this, it is not sufficient to just restore the original conditions and ignore the priorities for germ line repairs *vs* repairs to the soma - we are constrained by hysteresis.

It is known that some cells (such as T cells and neurons) exhibit this phenomenon of hysteresis: it takes a lower threshold to re-activate cells which have been previously activated. So germ line cells which have been previously 'activated' (*i.e.* made more responsive to repair inputs, and are repaired much easier compared to their somatic counterparts) are able to use any available repair resources and channel these to their advantage, whereas the somatic elements which are not easily responsive to the presence of repair resources, will need a much higher level of stimulation before they are activated. The threshold value of the repair mechanisms in germ line required to start the repairs is lower than the threshold value at which the somatic repair starts, if we assume that all other parameters involved in the process remain constant. This concept makes overcoming the barrier (b) in Fig. (**4**), more difficult.

An example of hysteresis in nature is based on the regime shift encountered in the transition of a lake from clear (low algae concentration) to murky (high algae concentration). Just like the case of soma-to-germ line resource allocation example, this is a 'path-dependent' process. An initial increase of the number of algae does not change the appearance of the water. If the number of algae increases, there will be a critical point (CP) when there is a 'phase transition' and the lake then turns turbid. An attempt to reverse this by reducing the number of algae will not be successful even if we reach the value of the original CP. Even if we continue reducing the number of algae well below the CP, the lake may still appear murky until a much lower level than the original CP is reached, when eventually it will clear. This phenomenon suggests that it would be difficult to 'undo' the existing situation where resources are allocated referentially to the germ line, because this process is already resilient following millions of years of evolution. Difficult, but it cannot be impossible. Other variables can be introduced in the system so that to facilitate the transition. One variable we can introduce is environmental challenges. However, applying environmental challenges alone without any specific direction, would not be sufficient to overcome the barrier, as we also need to consider the previously established natural history of the repair process in germ line. This is not to say that a reversal is impossible. Mathematically, the Bouc-Wen model of modelling hysteresis [28] predicts that hysteresis is versatile and may be modelled in bi-directional systems [29] (such as soma-to-germ line and germ line-to-soma).

Here we can introduce another variable that can be used in order to counteract the process of hysteresis: 'nudging.' The cybernetic concepts of 'stigmergy' and 'nudging' are quite relevant in this discussion. These concepts may be used in an integrated way in order to intentionally guide self-organisation of a system, perhaps, driving it out of an already existing path, and making it to operate along different, intentional and pre-planned lines. This discussion has been developed by Belgian cyberneticist Francis Heylighen, and he quotes (personal communication 1 January 2016):

Here are some thoughts on how the stigmergy and nudge paradigms can be integrated. First, both are ways of stimulating individual or collective action via the environment. In the case of stigmergy, a pattern of action self-organizes via the trace it leaves in the environment. Nudging, on the other hand, is an outside, controlled intervention to shape the environment in such a way that it promotes desirable actions (or discourages undesirable action). Together, they should be able to produce guided self-organization: nudges guide the action in the right direction, stigmergy lets the action develop and self-organize so as adapt to the circumstances.

The idea came to me when I was looking through a colouring book for adults. The principle of such books is to elicit artistic creativity (a beautiful pattern of colours) by already providing a "guide" (the black and white lines on the page) that makes it easy to get an aesthetic result and that stimulates the imagination. If the black and white drawing is fully specified, the creativity is, of course, quite limited. However, some pages of the book just offered a beginning of a drawing, thus stimulating the user not only to colour but also to continue the drawing. The incomplete drawing is a nudge, inciting people to make further drawings. The remaining blank page is a stigmergic medium: every bit of line or colouring added to that empty space incites the individual to add more lines or colours.

Now, you might wonder why you need the nudge, given that providing any pattern or constraint biases the further action in a certain direction, thus in principle reducing the variety of possible developments. However, anyone who has been teaching art, writing or other kinds of creative endeavour knows that starting with a "blank page" is more likely to inhibit than to stimulate exploration. "Fear of the blank page" is how writer's block is described in Dutch. It is one of the reasons why I insist that my students should start with an outline when first trying to write a paper. It is also the reason why Wikipedia uses "stub" pages, which as yet only contain a sentence or two about their subject, but which are as such more likely to incite people to develop the subject than an empty or absent page.

Our mind works as an associative or reactive system: it does not start thinking in a vacuum, like Descartes imagined or as the "brain-in-a-picture" picture of cognition supposes, but always in reaction to some previous thought or perception. The more concrete and detailed that stimulus, the richer the resulting train of thought. One of the reasons for the effectiveness of stigmergy is that the stimulus (trace) is constantly being enriched by the activity, thus stimulating ever further activity. But without any stimulus to start with, stigmergy may take a very long time before it gets going. That is why it is useful to start with a good nudge (or "stub" or template), i.e. one that suggests a way of acting that is likely to be fruitful.

Of course, the nudge should not only be stimulating, but leave enough room for extended exploration outside the initial trace. That is why the beginning of a drawing is a better example of guided self-organization than a traditional colouring book. In the case of a paper, a good start may be a generic outline, consisting of questions that apply to about any domain, such as: What is the context of the problem you are going to tackle? Why is it important to tackle that problem? Which other approaches have already been tried? Why aren't these sufficient? How is your approach different from these other approaches? ...

Let's think about how we can apply these ideas in the context of urban design, for instance focus on the use of open spaces to stimulate a fluid sense of self. Here, the emphasis is on promoting self-organization, but according to the above reasoning, you need more than an open space for that: this space should already contain some stimuli that suggest ways to use that space constructively (like the ping-pong table installed in a traffic-free street in Brussel), and that ideally enrich the pattern of use stigmergically. Another example is that of chairs left in a public park that tend to arrange themselves stigmergically, clustering around the most popular areas. The loose chairs are an initial nudge that says "bring me wherever you want to sit". But after the chairs have clustered, the self-organization comes to a halt, and further actions are merely local fluctuations, so creativity is limited. A possible combination of both approaches would be some kind of lego-like building blocks for urban structures (chairs, tables, benches, exercise infrastructure, playgrounds, sculptures, stalls...) that people could assemble and disassemble in different places and combinations. To nudge the process and inspire further creativity, some useful assemblies would already be provided from the start. This would create a stimulating, open space for exploration, while allowing stigmergic enrichment and self-organization. But it is not obvious how to design the initial building blocks, which should be sufficiently robust and versatile to allow many possible uses and abuses...

These notions are mentioned here in order to highlight the fact that the examples regarding cognitive stimulation (for instance, such as those in the Appendix) provide a general nudging towards the correct way to utilise our cognition, and then allow for our creativity to develop along individual preferences. However, it is also possible that the principle of nudging may apply in the context of forcing a re-allocation of resources from germ line to soma, as discussed above, and in other chapters. By nudging (*i.e.* increasing cognitive challenges and exposure to actionable information) we guide the system in a positive way and direct it towards a situation where it can self-organise in a manner that we desire, *i.e.* in withholding repair resources from the germ line and directing this back to the soma. In this way we see that cybernetic general principles such as stigmergy and nudging, can be applied in the biology of ageing and clarify how the system can be made to work to our advantage.

But returning to the issue of counteracting hysteresis (path dependency) we may also need to consider introducing another variable which may help shift the system out of existing and pre-determined paths. This variable can be termed 'lifestyle harmonisation'. This refers to a situation where an appropriate lifestyle is in accordance with the new technological environment: less emphasis on the physical, and more on the cognitive. This may involve less physical exercise and a more sedentary life **but only** in association with a sustained and frequent cognitive effort. This will be discussed below once I have dealt with some futurist issues.

A New Type of Human?

Several academics have speculated about the real or hypothetical emergence of a new human sub-species, which reflect the integration of biological humans with technology. Examples of terms used are:

1. Homo sapiens technologicus [30] and discussed here [31]:
2. Primo posthuman [32]
3. The Noeme [33]
4. Cyborg (and cyborg anthropology). The integration of organic function with technological function involving cybernetic mechanisms of feed-back and control [34, 35]
5. Metaman (Humanity and technology viewed as a global superorganism) [36]
6. Transhuman [37]

Irrespective of what term is used, it is becoming clear that humanity is now in the process of transiting into a new state which, for the first time in history, gradually and rapidly abandons existing 'biological-only' conventions and places more emphasis on a technological-biological symbiosis instead. We are increasingly being integrated with technology, including ambient intelligence. Whether we like it or not, we are being compelled to share information and become integral components of a wider entity which is the Global Brain [38]. This symbiosis is bound to have an impact on our biological systems. Our adaptation mechanisms are capable of absorbing a certain degree of integration, but we need to study the details of this process because it will help us modify or modulate it.

However, how can we conciliate the fact that while conventional (biological-only) advice about a healthy lifestyle and physical exercise is, without a doubt, still valid, and yet the new evolutionary paradigm suggests that we are becoming less physical and more cognitive instead? A technology-driven human will not need to be physically strong in order to survive, because technology will provide a range of options necessary to overcome physical limitations. The trade-offs between cognition and physical strength will thus be softened. Therefore we may now need to redefine the priorities of our lifestyle. A hitherto considered 'healthy' lifestyle may not be suitable to the new type of techno-biological human that is currently emerging, and it may now be advisable for some humans to shift the meaning of 'healthy' from the physical to the cognitive.

This is not to suggest that the need to exercise is totally eliminated but it may be more appropriate within a technological modern environment to make more effort to maintain cognition and intelligence (successful problem-solving), rather than aim to become physically stronger. Within a hormetic framework this is also a matter of balance: recreational exercise and relaxation have inherent values which cannot be overlooked at this stage. On the other hand, extreme or prolonged cognitive effort may result in information fatigue and overshoot the organism's adapting capabilities. Therefore, a balanced ideal should be found between excessive physical exercise and excessive mental challenges.

If during our evolution, say, we had to use our brain 20% and our physical activities 80% of the available time, this ratio would now be inappropriate in a society where there is more emphasis on the cognitive. As a general guidance and based on the broad concept of the Pareto distribution which is a power law probability distribution (https://en.wikipedia.org/wiki/Pareto_distribution), we may be able to adapt more successfully to a modern environment if we reverse this ratio: use our brain 80% and be physical 20% of our available time. This will still provide a suitable mix of physical and cognitive abilities, at least for a section of humanity. I am not arguing for a total replacement for exercise, but for a general propensity to exercise less and think more. This ratio is more appropriate in a word where cognition increasingly carries more value.

In my view it may be possible to account for the apparent conundrum – do we follow a mostly physical or a mostly cognitive challenging lifestyle? I have observed that humanity is not homogeneous with respect to cognitive priorities, intellectual expectations and creativity. This may mean that advice about a healthy lifestyle depends on what is considered 'health' by different groups of populations. In a technologically-driven world an appropriate definition of health is not the well-known one proposed by the World Health Organisation (for instance http://www.who.int/about/definition/en/print.html) but the one proposed in a British Medical Journal editorial: "The ability to **adapt** and self-manage in the face of social, physical, and emotional **challenges** (emphasis mine) [39]. Below I described a 'snap-shot' view where humans are divided, not by economic or political priorities, but by cognitive ones (Fig. **5**).

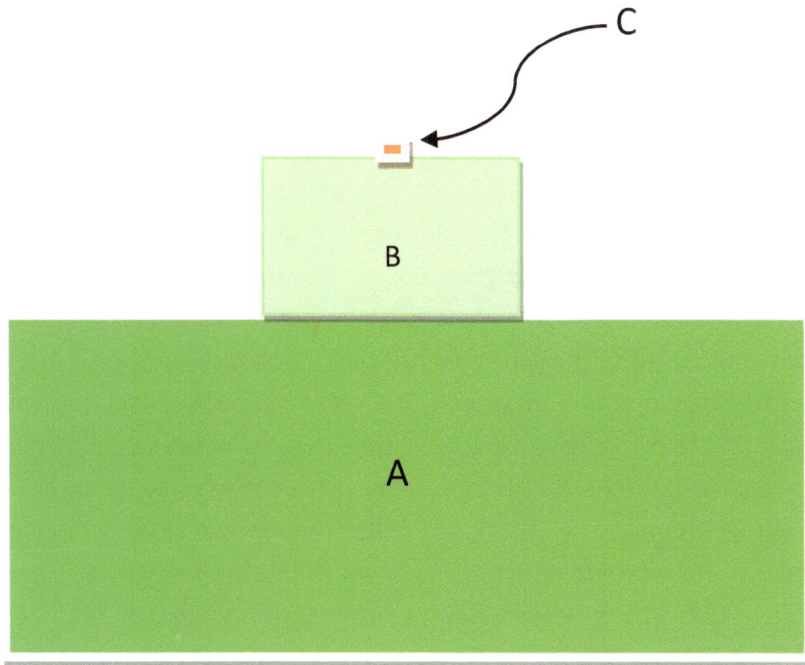

Fig. (5). The cognitive division of humanity: three broad groups. The definitions of A, B and C group characteristics are given in the text.

Group A: This encompasses the great majority of humanity, mostly not 'self-actualised' [40] individuals, the conventional, those who have poor imagination and no motivation to achieve a nobler or a greater aim. They are those who live their everyday life (irrespectively of being rich or poor), being contented with what they have, happy to have a family, look forward to their retirement and die when their time comes. These are the *r-strategists* [41], with shorter lifespans and high rates of reproduction (more than two children per couple). These people are essential in channelling energy towards higher levels, and some of their progeny may be able to move to group B. They are likely to benefit from traditional health advice, such as moderate physical exercise, generally healthy diet and lifestyle, simple dietary supplements, nutritional aids, avoiding excesses *etc*. As these people contribute mostly passively to the evolutionary process (and do not play any part in purposefully guiding evolution), they continue to be subjected to the slow process of natural selection, and remain constrained by the predictions of the disposable soma theory, being disposable/expendable, with more emphasis on the physical rather than the cognitive.

Group B: Perhaps numbering in the order of one billion, these are the 'Motivated Intermediates', those who realise that there is more to life than just living and then dying. The 'semi-unconventional', with a fair degree of (perhaps misplaced) sophistication and inapt creativity, they believe that humans must achieve something nobler than mere survival. They are not quietly fulfilled with everyday life and think that they should strive for something higher. They feed energy to the higher level, and some of their progeny may be able to move to group C. They experience a longer lifespan, are able, in principle, to reach super-centenarian and have lower numbers of offspring (two or fewer children per couple). They can benefit from semi-unconventional health advice, such as mostly what we have presented in the initial chapters of the book: Paleo and power law lifestyle, 'Primal Blueprint' style activities (https://www.primalblueprint.com/) and 'Movenat' principles [42]. The inherent value of this group is that it prepares increasingly more members of humanity for entry into group C. Group B is a stage where people are primed, in a Dantesque purgatorial sense, for a philosophically superior, greater and nobler stage.

Group C: Population approximately 50-100 million. Not necessarily the richest financially, but those with high levels of intellectual sophistication, persistent motivation, imagination, vision and creativity. The free-thinking liberals, the illuminated, the *K-strategists* [41] with long (most likely to eventually experience indefinite) lifespans and low or absent reproduction (nil children per couple-no couples but mixed sexuality). They are more prone to follow unconventional health advice with less emphasis on attaining full physical fitness and more emphasis on expanding cognitive qualities. They are <u>meaningfully</u> hyper-connected, and fully compliant with the Law of requisite Usefulness (which states that the length of retention of an agent within a network is proportional to that agent's contribution to the overall adaptability of the network) [43]. They believe that humans are not here just to experience as much personal pleasure as possible before they die, but have developed the ability (even, the duty) to constructively change their environment, and are taking active steps in order to make a difference in the world. They are the intentional evolutionaries (those who deliberately and intelligently manipulate our evolutionary process) [44]. These people are much less expendable and more prone to be subjected to the Indispensable Soma hypothesis [45] being very costly to replace and less so to repair in the biological sense (so they are more likely to be retained by evolutionary necessities).

An Unconventional Worldview: Harmonising our Lifestyle

The current well-accepted health advice is to live a 'natural' lifestyle, eat natural food, and exercise regularly. But this kind of lifestyle was appropriate for humans who depended on hunting, gathering, working the land, finding their way in the forest and having to be physically strong in order to cope with the rigours of life. However, our environment is now different from that of our ancestors. Now some sections of humanity are becoming very dependent on technology. Our cognitive processes need instant energy from foods not normally considered healthy (such as sugars) [46]. It is no longer evolutionarily relevant to be able to run marathons or exercise at the gym when, instead, we can contribute better to humanity by sharing actionable and meaningful information which may enlightened others. The information burden carried by modern technology (Ambient Intelligence, the Global Brain, The Internet of Things) is having unprecedented impact on our biological mechanisms, which is at odds with commonly-held beliefs about what is natural [47].

One may argue that our biological heritage still depends on previously defined parameters and that it would be difficult to change this dependency. However, this is not valid as a concept. Fast epigenetic changes and other biological mechanisms based on degeneracy, exaptation and resilience discussed above, are perfectly able to compensate for this new techno-cultural lifestyle. This is not to suggest that the need to exercise is totally eliminated but it may be more appropriate within a technological modern environment to make more effort to maintain cognition and intelligence (successful problem-solving), rather than aim to become physically stronger. I will discuss further details and some research evidence regarding this topic in the next chapter. However, much more research is needed in order to grasp a clear understanding of these processes. Current research is biased in looking for the cognitive benefits of physical activities. There is little research examining the physical effects of cognitive activities.

Another crucial mind-set that needs to change is the one which relies on rejuvenation biotechnologies as a way for achieving a negligible rate of ageing (a minimal mortality as a function of age [48]. It would be a great leap of faith for someone to realise that, the elimination of age-related degeneration (and thus radically extending the human lifespan) will not, in fact, be achieved by using a physical intervention [49, 50]. In my view, it would only be achieved through cultivating the transition from groups A and B to the group C as described above [51].

These arguments concerning a technology-driven change in conventional worldviews will be explored in greater detail in the last chapter of this book.

CONCLUSION

Epigenetic regulation mediates adaptation to the environment, and when this environment is of a specific type (an information-sharing, fast changing techno-cultural setting) then fast adaptation to this may have an impact upon life-extending biological processes. This epigenetic regulation, together with genetic and cell-specific factors can influence longevity and has been studied within a non-reductionist, integrative framework of ageing [52]. The reductionist view that ageing can be manipulated by conventional lifestyle changes and simple biomedical repairs is unlikely to lead to any appreciable practical results that can be used in order to diminish the impact of age-related degeneration. Instead, a wider approach more likely to succeed would be to study the role of complexity, regulation and adaptation, and harness the function of fundamental processes such as the role of emergence and the 'direction' of human evolution [22]. This world-view has been encapsulated in philosophical reflections which transcend time and place, such as *"Tempora mutantur, nos et mutamur in illis"* ("Times change, and we change with them"). As our environment, society and culture change, so must we, in order to adapt to the change and increase our fitness and chance of survival [53]. One way we can effect a positive change is to increase exposure to meaningful technological information that requires action [54] and exploit the nature of relatively fast epigenetic mechanisms which, bound by evolutionary constraints, must result in suitable beneficial adaptations, and thus extended survival.

REFERENCES

[1]   Kazantzakis N. The Saviors of God: Spiritual Exercises. New York: Touchstone / Simon and Schuster 1960.

[2]   Schejter A, Agassi J. On the Definition of Life. J Gen Philos Sci 1994; 25(1): 97-106.
[http://dx.doi.org/10.1007/BF00769279]

[3]   Frieden BR, Gatenby RA. Information dynamics in living systems: prokaryotes, eukaryotes, and cancer. PLoS One 2011; 6(7): e22085.
[http://dx.doi.org/10.1371/journal.pone.0022085] [PMID: 21818295]

[4]   Zaman L, Meyer JR, Devangam S, Bryson DM, Lenski RE, Ofria C. Coevolution drives the emergence of complex traits and promotes evolvability. PLoS Biol 2014; 12(12): e1002023.
[http://dx.doi.org/10.1371/journal.pbio.1002023] [PMID: 25514332]

[5]   Stewart JE. The direction of evolution: the rise of cooperative organization. Biosystems 2014; 123: 27-36.
[http://dx.doi.org/10.1016/j.biosystems.2014.05.006] [PMID: 24887200]

[6]   Beigi S. Minfulness Engineering. A Unifying Theory of Resilience for the Volatile, Uncertain, Complex, and Ambiguous (VUCA) World. PhD Thesis,. University of Bristol UK. Available from: https://vub.academia.edu/ShimaBeigi/Thesis-Chapters 2015.

[7]   Tlaskalova-Hogenova H, Tuckova L, Mestecky J, *et al.* Interaction of mucosal microbiota with the innate immune system. Scand J Immunol 2005; 62(1): 106-13.
[http://dx.doi.org/10.1111/j.1365-3083.2005.01618.x]

[8]   Belkaid Y, Segre JA. Dialogue between skin microbiota and immunity. Science 2014; 346(6212): 954-9.
[http://dx.doi.org/10.1126/science.1260144] [PMID: 25414304]

[9]   Fuhrmann A, Engler AJ. The cytoskeleton regulates cell attachment strength. Biophys J 2015; 109(1): 57-65.
[http://dx.doi.org/10.1016/j.bpj.2015.06.003] [PMID: 26153702]

[10]   Chow BW, Gu C. The molecular constituents of the blood-brain barrier. Trends Neurosci 2015; 38(10): 598-608.
[http://dx.doi.org/10.1016/j.tins.2015.08.003] [PMID: 26442694]

[11]   Keenan MJ, Marco ML, Ingram DK, Martin RJ. Improving healthspan *via* changes in gut microbiota and fermentation. Age (Dordr) 2015; 37(5): 98.
[http://dx.doi.org/10.1007/s11357-015-9817-6] [PMID: 26371059]

[12]   Bartel J, Krumsiek J, Schramm K, *et al.* The human blood metabolome-transcriptome interface. PLoS Genet 2015; 11(6): e1005274.
[http://dx.doi.org/10.1371/journal.pgen.1005274] [PMID: 26086077]

[13]   Vacher G, Niculita-Hirzel H, Roger T. Immune responses to airborne fungi and non-invasive airway diseases. Semin Immunopathol 2015; 37(2): 83-96.
[http://dx.doi.org/10.1007/s00281-014-0471-3] [PMID: 25502371]

[14]   Graziottin A. Vaginal biological and sexual health--the unmet needs. Climacteric 2015; 18 (Suppl. 1): 9-12.
[http://dx.doi.org/10.3109/13697137.2015.1079408] [PMID: 26366794]

[15]   Barve A, Wagner A. A latent capacity for evolutionary innovation through exaptation in metabolic systems. Nature 2013; 500(7461): 203-6.
[http://dx.doi.org/10.1038/nature12301] [PMID: 23851393]

[16]   Maleszka R, Mason PH, Barron AB. Epigenomics and the concept of degeneracy in biological systems. Brief Funct Genomics 2014; 13(3): 191-202.
[http://dx.doi.org/10.1093/bfgp/elt050] [PMID: 24335757]

[17]   Hou YM, Gamper H, Yang W. Post-transcriptional modifications to tRNA--a response to the genetic code degeneracy. RNA 2015; 21(4): 642-4.
[http://dx.doi.org/10.1261/rna.049825.115] [PMID: 25780173]

[18]   Mason PH. Degeneracy: Demystifying and destigmatizing a core concept in systems biology. Complexity 2015; 20: 12-21.
[http://dx.doi.org/10.1002/cplx.21534]

[19]   Tononi G, Sporns O, Edelman GM. Measures of degeneracy and redundancy in biological networks. Proc Natl Acad Sci USA 1999; 96(6): 3257-62.
[http://dx.doi.org/10.1073/pnas.96.6.3257] [PMID: 10077671]

[20]   Kitano H. Towards a theory of biological robustness. Mol Syst Biol 2007; 3: 137.
[http://dx.doi.org/10.1038/msb4100179] [PMID: 17882156]

[21]   Rattan SI. Hormesis in aging. Ageing Res Rev 2008; 7(1): 63-78.
[http://dx.doi.org/10.1016/j.arr.2007.03.002] [PMID: 17964227]

[22]   Kyriazis M. Editorial: Novel approaches to an old problem: insights, theory and practice for eliminating aging. Curr Aging Sci 2014; 7(1): 1-2.
[http://dx.doi.org/10.2174/1874609807011140703103943] [PMID: 25056407]

[23]   Carlson JM, Doyle J. Highly optimized tolerance: robustness and design in complex systems. Phys Rev Lett 2000; 84(11): 2529-32.
[http://dx.doi.org/10.1103/PhysRevLett.84.2529] [PMID: 11018927]

[24]   Szalay MS, Kovács IA, Korcsmáros T, Böde C, Csermely P. Stress-induced rearrangements of cellular networks: Consequences for protection and drug design. FEBS Lett 2007; 581(19): 3675-80.
[http://dx.doi.org/10.1016/j.febslet.2007.03.083] [PMID: 17433306]

[25]   Flatt T, Amdam GV, Kirkwood TB, Omholt SW. Life-history evolution and the polyphenic regulation of somatic maintenance and survival. Q Rev Biol 2013; 88(3): 185-218.
[http://dx.doi.org/10.1086/671484] [PMID: 24053071]

[26]   Calvanese V, Lara E, Kahn A, Fraga MF. The role of epigenetics in aging and age-related diseases. Ageing Res Rev 2009; 8(4): 268-76.
[http://dx.doi.org/10.1016/j.arr.2009.03.004] [PMID: 19716530]

[27]   Calvanese V, Lara E, Kahn A, Fraga MF. The role of epigenetics in aging and age-related diseases. Ageing Res Rev 2009; 8(4): 268-76.
[http://dx.doi.org/10.1016/j.arr.2009.03.004] [PMID: 19716530]

[28]   Bouc R. Forced vibration of mechanical systems with hysteresis.

[29]   Wen YK. Method for random vibration of hysteretic systems. J Eng Mech 1976; 102(2): 249-63.

[30]   Gingras Y. Éloge de l'homo techno-logicus. Saint-Laurent, Québec: Les Editions Fides 2005.

[31]   Zehr PE. Future Think: Cautiously Optimistic About Augmenting the Human Brain. Front Syst Neurosci 2015; 9: 72.
[http://dx.doi.org/10.3389/fnsys.2015.00072] [PMID: 26042003]

[32]   Vita More N. http://www.kurzweilai.net/radical-body-design-primo-posthuman 2015.

[33]   Kyriazis M. Technological Integration and hyperconnectivity: Tools for promoting extreme human lifespans. Complexity 2015; 20(6): 15-24.
[http://dx.doi.org/10.1002/cplx.21626]

[34]   Schermer M. [A cyborg is only human]. Ned Tijdschr Geneeskd 2013; 157(51): A6879.
[PMID: 24345361]

[35]   Giselbrecht S, Rapp BE, Niemeyer CM. The chemistry of cyborgs--interfacing technical devices with organisms. Angew Chem Int Ed Engl 2013; 52(52): 13942-57.
[http://dx.doi.org/10.1002/anie.201307495] [PMID: 24288270]

[36]   Stock G. Metaman: The Merging of Humans and Machines into a Global Superorganism. New York: Simon & Schuster 1993.

[37]   More M. Transhumanism: Towards a Futurist Philosophy. Available from: https://web.archive.org/web/20051029125153/http://www.maxmore.com:80/transhum.htm 1990.

[38]   Last C. Human Metasystem Transition Theory (HMST). J Evol Technol 2015; 25: 1-16.

[39]   Godlee F. What is health? BMJ 2011; 343: d4817.
[http://dx.doi.org/10.1136/bmj.d4817]

[40]   Mark. E. Rediscovering the Later Version of Maslow's Hierarchy of Needs: Self-Transcendence and Opportunities for Theory, Research, and Unification. Rev Gen Psychol 2006; 10(4): 302-17.
[http://dx.doi.org/10.1037/1089-2680.10.4.302]

[41]   Heylighen F, Bernheim JL. From Quantity to Quality of Life: r-K selection and human development. Available from: http://pespmc1.vub.ac.be/papers/r-kselectionqol.pdf. 2015.

[42]   Le Corre E. MovNat. Auberry, United States: Victory Belt Publishing 2013.

[43]   Kyriazis M. The law of requisite usefulness. Available from: http://scienceblog.com/76141/law-requisite-usefulness/#isdAh6Dz3dX6duP0.97. 2015.

[44]   Stewart J. Evolution's Arrow: the direction of evolution and the future of humanity. Canberra: The Chapman Press 2000.

[45]   Kyriazis M. The Indispensable Soma hypothesis. Available from: http://scienceblog.com/79258/indispensable-soma-theory-ageing/#Fm14ig7gzuQtUD8H.97. 2015.

[46]   Kann O, Papageorgiou IE, Draguhn A. Highly energized inhibitory interneurons are a central element for information processing in cortical networks. J Cereb Blood Flow Metab 2014; 34(8): 1270-82.
[http://dx.doi.org/10.1038/jcbfm.2014.104] [PMID: 24896567]

[47]   Kyriazis M. Systems neuroscience in focus: from the human brain to the global brain? Front Syst Neurosci 2015; 9(7) eCollection
[http://dx.doi.org/10.3389/fnsys.2015.00007]

[48]   Zealley B, de Grey AD. Strategies for engineered negligible senescence. Gerontology 2013; 59(2): 183-9.
[http://dx.doi.org/10.1159/000342197] [PMID: 23037635]

[49]   Kyriazis M. The impracticality of biomedical rejuvenation therapies: translational and pharmacological barriers. Rejuvenation Res 2014; 17(4): 390-6.
[http://dx.doi.org/10.1089/rej.2014.1588] [PMID: 25072550]

[50]   Kyriazis M. Translating laboratory anti-aging biotechnology into applied clinical practice: Problems and obstacles. World J Transl Med 2015; 4(2): 51-4.
[http://dx.doi.org/10.5528/wjtm.v4.i2.51]

[51]   Kyriazis M. A cognitive-cultural segregation and the three stages of aging. Curr Aging Sci 2015. [Epub ahead of print].
[PMID: 26651458]

[52]   Nikoletopoulou V, Kyriakakis E, Tavernarakis N. Cellular and molecular longevity pathways: the old and the new. Trends Endocrinol Metab 2014; 25(4): 212-23.
[http://dx.doi.org/10.1016/j.tem.2013.12.003] [PMID: 24388148]

[53]   More M, Vita-More N. The Transhumanist Reader: Classical and Contemporary Essays on the Science, Technology, and Philosophy of the Human Future. Chichester, UK: John Wiley & Sons 2013.
[http://dx.doi.org/10.1002/9781118555927]

[54]   Kazantzakis N. The Saviors of God: Spiritual Exercises Friar K (translator). New York: Touchstone / Simon and Schuster 1960.

Frontiers in Aging Sciences, 2016, *Vol. 1*, 169-200   **169**

CHAPTER 7

Energy, Entropy and Complexity: Thermodynamic and Information-Theoretic Perspectives on Ageing

Atanu Chatterjee

Indian Institute of Technology Madras, Chennai-600036, T.N., India

Abstract: The human body is a complex system. It has a natural ability to grow and develop over time, as well as adapt to the frequent changes in the surroundings. The complexity associated with it and the various processes it undergoes lie in its structure and functionality. Complex chemical reactions play a central role in the evolution of structure and organization of this complex adaptive system, which are inherently directed along the paths of maximum entropy production. Entropy production causes a system to degrade itself by the gradual consumption of free-energy to a more thermodynamically-stable state. However, the human body, and open systems in general, have a tendency to preserve or increase order and complexity with time. This phenomenon of spontaneous appearance of order is known as self-organization. Thus, the human body has a structure and several underlying functions that give rise to organization. Simultaneously, it prevents the destruction of this state of organization by self-organizing itself with time. In the following chapter, we will look into the details of (self) organization from a physical and an information perspective, in order to see the relationship between complexity and the growth in organization of a system with time. Further, understanding the concept of (self) organization from a functional perspective is also important as it will allow us to relate metabolic reaction-sets to structural symmetry, metabolites to interacting nodes, and mapping these interactions into complex networks. Finally, our aim will be to relate ageing to the loss of energy, information, organization and functionality of the human body with time.

Keywords: Complexity, Dissipative Systems, Principle of Least Action, Second Law of Thermodynamics.

INTRODUCTION

The human body, its structure and functionality have always been a topic of extreme interest in the scientific community. The structural complexity exhibited by the human body has always intrigued mankind. How do small entities such as the cell, metabolites and protein molecules, which are invisible to the naked eye, amalgamate to form bigger molecules, tissues and organs is a very interesting question. In general, the concept of life is fascinating. One of the most important scientific accounts on this fascinating subject can be found in Schrödinger's book which dealt with the definition of life. In this book, which was based on a course of public lectures that he had delivered in 1943, he had focused primarily on one important question: *"How can the events in space and time which take place within the spatial boundary of a living organism be accounted for by physics and chemistry?"* [1].

In this book, he speculated about an 'aperiodic crystal' that contained the genetic information in its configuration. Although the existence of DNA was established back in 1863, its specific purpose in reproduction still remained a mystery. By 1903 people thought that chromosomes were the heredity units but it wasn't until 1933 that Jean Brachet showed that the chromosomes were composed of DNA. By 1944 there was experimental evidence for DNA being the heredity material. The ideas presented by Schrödinger in his book provided an early theoretical description of how the storage of genetic information would work [2]. Although there are accounts of ideas on the physical basis of life even before Schrödinger, such as Muller's article on "Variation due to change in the individual gene", Schrödinger's approach to the problem is particularly interesting[1]. Primarily, Schrödinger did not discriminate between the processes occurring outside the cell with those occurring inside. Secondly, his treatment of the problem was based on the fundamental laws of physics or from the first principles [4, 5]. He identified the field of thermodynamics and statistical mechanics to provide himself with an answer. The Second Law of Thermodynamics is a powerful tool in the hands of a physicist [6 - 11]. Schrödinger observed that the purpose of life is to increase the complexity and the information content in a living being, whereas, we being surrounded by the Second Law should observe just the opposite. This led him to coin the term, *negative entropy* or *negentropy* – the entropy that a living system exports into the surroundings in order to keep its entropy low, which he corrected in the later editions by stating that life feeds on free energy [1]. The other interesting observation that Schrödinger made was that most physical laws at macro scales are a result of chaos at the finer levels. This realization of Schrödinger's is what we know as the phenomenon of self-organization [12, 13]. Thus, Schrödinger in his explanation of life provided us with two important insights – his formulation of life as a thermodynamic process and his observation on the appearance of global order from local fluctuations. Both the observations lead us to visualize the human body as a complex adaptive system [12, 14].

The domain of Complex Adaptive Systems (CAS) is a relatively new field when compared to its counterparts, such as classical thermodynamics or statistical mechanics. CAS are macroscopically complex, made up of numerous interacting finer entities, and are adaptive to the fluctuations in the surrounding environment. Typical examples of CAS include the human body, the cell, the stock exchange, the ant colony, *etc.* It is observed that the property of these systems as a whole differ from each of their individual agents, which is contrary to the usual stance of classical physics. These systems are adaptive and complex, yet they lack a central coordination. The entities that make up a CAS often follow simple rules to interact with each other. In Fig. (**1**), we can observe that the left extremity of the bell-curve is occupied by causal-Newtonian physics, according to which, the path of each interacting particle in a system is deterministic. In the extreme right of the bell-curve in Fig. (**1**), we observe dis-ordered complexity arising due to stochasticity in the system. In the middle of the curve, lies ordered complexity, such as the cell or the human body, which although complex have structure and organization. In ant colonies, for example, the only mode of interaction between the ants is through the pheromone trail that an ant leaves while foraging for food. The simple idea of 'following the trail' leads ants to find the shortest path in the real world between the colony and the food source [16]. Also, in the organizational hierarchy of the ant colony, there does not exist any central control, rather the control between various agents is distributed, and the above simple rule gives rise to a global coordination in the entire colony. Interestingly, the numerous interactions and distributed coordination makes the system robust to perturbations in the surroundings [12]. Similar to the above example of an ant colony is the human body, the brain or one of the smallest building units of the human body, the cell. All of these are made of finer entities which interact with each other, giving rise to emerging patterns, thus making the system as a whole, robust and homeostatic. One of the earliest examples of the emergence of patterns due to self-organization is the Rayleigh-Bernard convection cell [17]. Rayleigh-Bernard convection is the coordinated movement of a fluid confined between two thermally conducting plates. When one of the plates is heated from the bottom, the temperature difference causes a random movement of fluid particles allowing energy transfer between the plates. However, above a certain temperature gradient, this random motion transforms into a coordinated movement, which gives rise to the emergent patterns of convection rolls (see Fig. **2**) and allows a much more efficient energy dispersal between the plates.

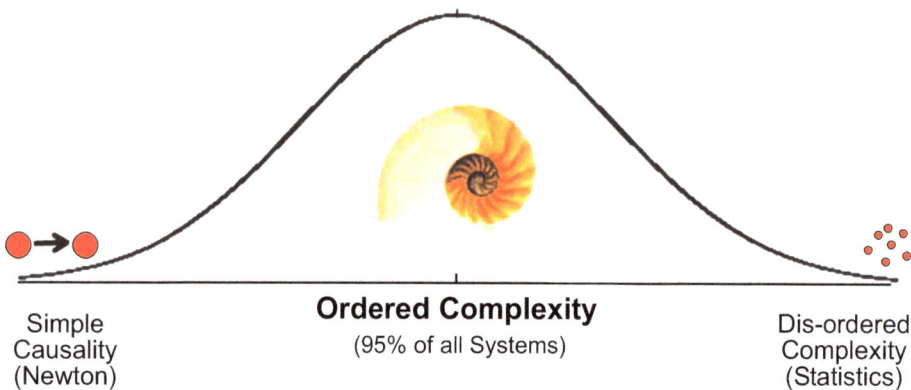

Fig. (1). Figure shows the distribution of complexity in nature. Extreme left of the curve denotes Newtonian universe with deterministic laws, whereas extreme right denotes the domain of dis-ordered complexity or statistical mechanics. The significant middle region of the curve captures the ordered complexity or those complex systems which exhibit a structure with a definite function [15].

The general struggle for existence of animate beings is not a struggle for raw materials – these, for organisms, are air, water and soil, all abundantly available – nor for energy which exists in plenty in anybody in the form of heat, but a struggle for (negative) entropy, which becomes available through the transition of energy from the hot sun to the cold earth – Boltzmann [18]. As noted by Schrödinger and also in our discussion of the Bernard-Rayleigh convection cells, energy or more specifically, energy dispersal, plays a key role. The role of entropy in life is often discussed in many perspectives. Since energy is the most important requirement for sustenance and survival, the human body can be considered as a thermodynamic system operating between the energy gradients of source and the surrounding environment. The energy export between the human body and the source-surrounding pair can be understood as the entropy which is dimensionally the energy (in form of heat) exchanged by the system with the surrounding at a given temperature. To be specific, the energy in the context of life is better understood as the free energy content, particularly, Gibb's free energy[2]. It is argued that living organisms preserve their internal order by consuming free energy from their surroundings in the form of nutrients and sunlight, and returning an equal amount of energy as heat (and entropy) to their surroundings [20]. The increase in entropy during a process is captured by the change in Gibb's free energy. This definition of entropy as energy dispersal from a state of high free energy to a state of low free energy is extremely important in the context of complex systems. Also, note that the presence of a gradient will *spontaneously* drive a system from a higher energy state to a lower energy state much like diffusion, where concentration gradients play the role of a driving force [19, 21, 22]]. Although this explanation looks very simple at first, there are a couple of questions surrounding its validity. When a single valid law can explain life, why do we have such diversity in nature? Why the complexity of living systems increased over time, as has can be seen from the organizational hierarchy? The Second Law of Thermodynamics explicitly states that the entropy of an isolated system will increase with time, however, living systems are all but isolated and are systems which are at a continuous state of non-equilibrium. So, how do these systems, which are far from equilibrium, fit into the Second Law paradigm?

To answer each of the above questions, we need to look at the multiple facets of the general problem, *i.e.*, to understand the structure and working of a complex adaptive system. The energy-entropic perspective will allow us to reason the cause of diversity in nature. We will see that the Second Law is not sufficient enough to describe why nature (or evolution) chooses one trajectory over the other during the course of any phenomenon. We will have to invoke other fundamental laws to understand the organizational structure and hierarchy in living systems. The visualization of complexity is tricky in this sense, as complexity in nature arises due to either of these two reasons: (**i**) multiple multi-scale interactions between the agents in a system, or (**ii**) due to indeterminism or lack of information of the precise behaviour of each agent in the system. In this chapter, we will not only look into the thermodynamic perspective of ageing from the context of free-energy, but we will also consider the complexity and information-theoretic aspects of it.

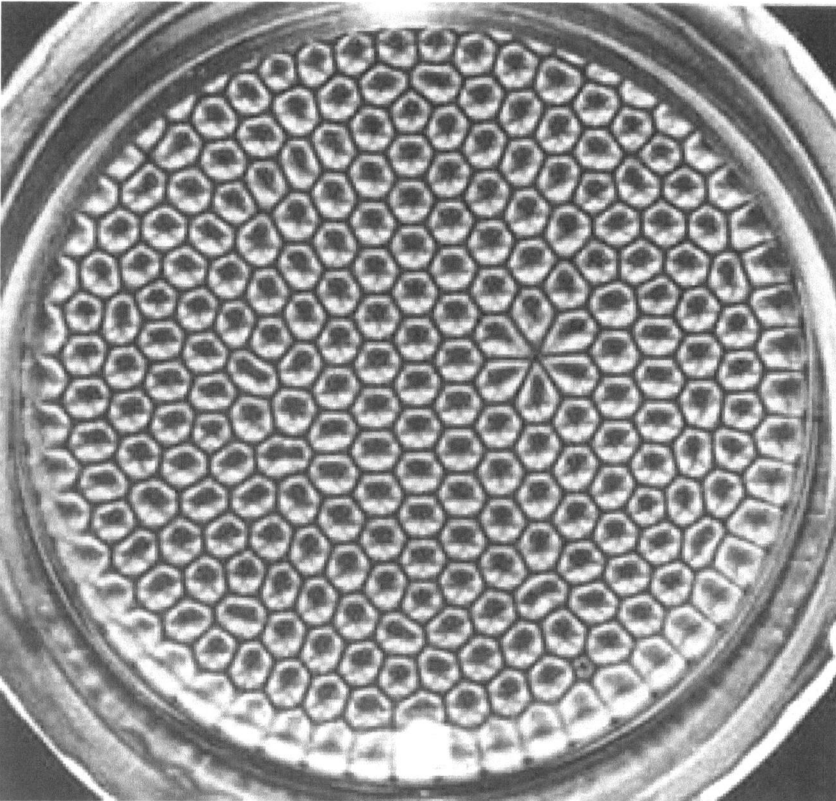

Fig. (2).   Figure shows the Rayleigh-Bernard convection cell. Observe the emergence of patterns of hexagonal cells which result in efficient energy transfer between the two thermally conducting plates [17].

Energy, Entropy and Senescence

The two reasons for the failure of classical physics in accurately describing real-world systems and phenomena lie in: **(i)** determinism and **(ii)** reductionism. Ironically, both of the above features are also considered as the achievements of Newtonian physics. Real world systems and processes are mainly indeterministic, as an element of uncertainty is always involved with every outcome. Uncertainty or stochasticity in the macroscopic world arises due to the presence of nonlinearity and a number of hidden (indeterministic) relationships between the variables controlling a process. Whereas, uncertainty in the microscopic world can be attributed to the quantum limitations of energy, dimension and measurements. Even though the direction of any natural process or its trajectory is not known beforehand, we can predict the most probable direction of its occurrence based upon energy considerations. One of the most fundamental contributions of classical physics is the Principle of Least Action, also known as Hertz's Principle of Least Curvature, Gauss's Principle of Least Constraint, Fermat's Principle of Least Time, or commonly, the Least Action Principle [23 - 31]. In simple language, the Least Action Principle[3]states that out of all possible paths to reach state **2** from state **1** (see Fig. **3**), only that particular path is chosen which minimizes action, or the path which provides least constraint, or along which the time taken is minimum. Similar to the Entropy Principle, which deals with natural processes by associating time-irreversibility in the sequence of events, the Action Principle gives an insight into the most probable direction for those sequences of events to progress. Be it the path taken by a refracting beam of light, a projectile in the sky or the wave-like motion of the electrons in a Hydrogen atom, the entire physics from relativistic mechanics to quantum mechanics can be derived from this fundamental law of nature. The real challenge, therefore lies in asking ourselves how these two fundamental laws of nature – the Least Action Principle and the Entropy Principle – are connected to each other.

Physically, action has the dimension of energy multiplied by time, which when expressed mathematically, is the product of change in energy (experienced by a particle along a path) multiplied by the time the particle takes to reach the final state from the initial (Hamilton's formulation), or the product of change in momentum experienced by a particle while traversing along a path times the length of the path (Maupertuis' formulation) [23]. The action as a metric has a versatile role in understanding the structure, property and organization in a physical system. Simply put, action can be described as the smallest metric of energy dissipation in a physical system arising out of interaction between the agents in a system, in terms of quanta, or in a networked system as the least unit of energy dissipation while traversing from one node to the other [32 - 38]. Also, action can be understood as a property of the system, a measure to define the functionality of a system and its response to the fluctuating surroundings, or when it undergoes a transition from one state to the other [39 - 47]. In reference to our previous remark on the connectedness of the Action Principle with the Entropy Principle, it would be worthwhile to look at the Second Law from both macro as well as microscopic perspectives. From the microscopic perspective, the Second Law can be reformulated as a statistical measure to quantify the disorder of an ensemble. Thus, the notion of entropy can also be broken down into a probabilistic framework, where the Second Law directs which next microstate (or macrostate) shall be occupied. Since, we have established a common ground, *i.e.*, probability to formulate both the Action Principle (as the most probable direction) and the Entropy Principle (as the most probable state), it is imperative that a strong connection exists between the two. The Second Law when rephrased, by taking into account the Least Action Principle, turns out to be a powerful tool to explain chemical reactions, diversity and evolution, or in general, the definition of life. The Action Principle presents before nature the most probable trajectory amongst all the existing (possible) ones for any process (natural, unconstrained, spontaneous) to occur (as action is minimized along that path). On the other hand, the Entropy Principle dictates the direction of the process – a unification of both will result into a selection of those states among the probable ones which reduce action. In terms of energy and time, the above statement can be reiterated as, "those possible states are selected along which free energy degradation (Entropy Principle) occurs in the least possible time (Least Action Principle)" [32, 48, 49].

The use of probability to understand the unified notion of entropy and action helps us in identifying the Second Law as a force that directs the dispersal of energy across gradients along least action paths [40, 50]. From a macroscopic perspective, the new formulation of the Second Law states that energy dispersal between a system and its surroundings will occur along the paths of least action, or the energy gradients shall be levelled in the least possible time. This can be used not only to understand the biological process of evolution and ageing, but also socio-economic processes like evolution of economies or societies. *Evolution is not a random sequence of events. Processes as flows of energy will themselves search by variation and select naturally those ways and means, such as species and societies or gadgets and galaxies that will consume free energy in the least time. In this way systems step from one state of symmetry to another by either acquiring or expelling at least one quantum of action. It is the photon, the basic building block of everything. A step down in free energy is an irreversible step forward in time[4].*

Multiple Trajectories

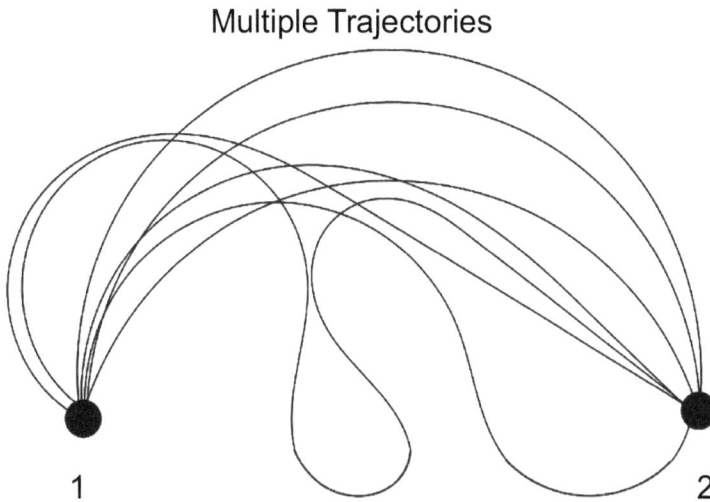

Fig. (3). Figure shows the numerous possible trajectories connecting state 1 and state 2. Out of all the possible trajectories the one with the least action is *naturally* selected. Along each of the paths the energy (the Hamiltonian) is conserved.

Although we incorporated action into our restatement of the Second Law, there are a couple of points that we need to address. We claimed that entropy can be visualized as dispersal of energy directed towards equalizing gradients, yet processes in nature do not readily let systems disperse energy, even when a gradient exists between the system and the surrounding[5]. More precisely, what is the force that binds a system and its interacting agents together, even though the system is thermodynamically 'open' (or dissipative)? An open system (all physical systems in nature are open systems) is always exposed to fluctuations from the environment due to changes in the energy landscape of the surrounding media (see Fig. **4**). If the purpose of life was to degrade free energy in the fastest possible way, then all lifeforms on the surface of the earth would have ceased to exist as soon as they appeared, instead of which we see diversity all around us. Thus, there is more to understand in the definition of entropy than energy dispersal alone. Also, there is more to a definition of a system than defining it simply as a collection of interacting agents [47, 51].

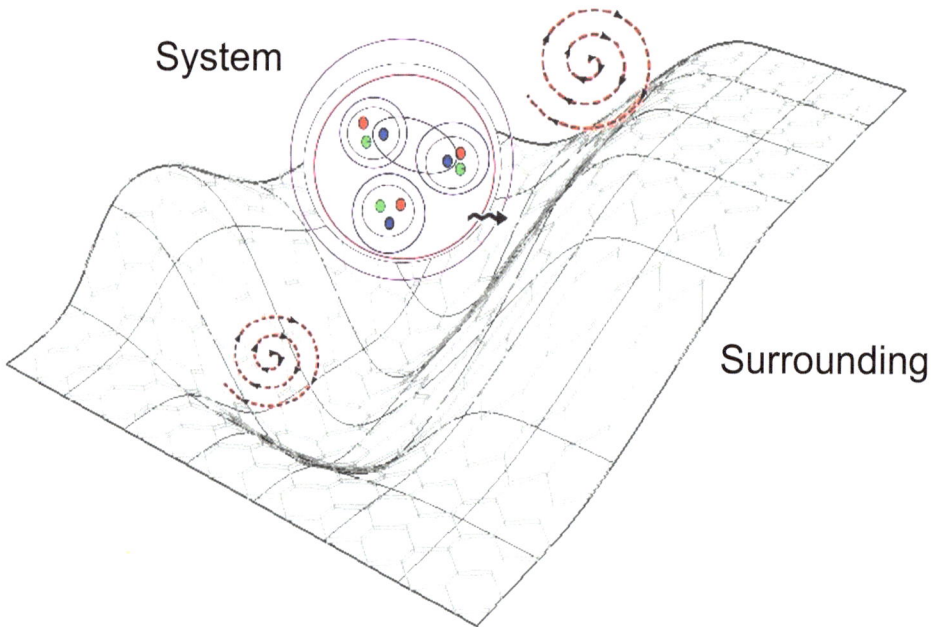

Fig. (4). Figure shows a system with interacting agents and a boundary separating the system from the surroundings. The figure also shows energy exchange between the system and the surrounding, where the surrounding is being represented as an energy landscape. The surrounding media is made up of photons (therefore units of action, in red spirals); and the continuous exchange of photons between the system (which is thermodynamically open) and the surroundings always tend to drive the system away from the state of equilibrium.

Thermodynamically, a system is that part of the universe in which we are interested in and which is separated from the rest of the universe by a boundary. The Laws of Thermodynamics implicitly assume that a boundary, a layer, exists between the system and the surroundings. All transfers of energy, entropy and mass are considered to take place when these quantities precisely cross the system boundary. Thus, a system boundary gives a structure to the system, and also acts as a separating layer between the surrounding environment and the system's internal machinery. A system can also be defined as a collection of (micro-) states, agents or particles, which can be extended to the definition of action (in terms of interactions). The interactions between system components in the form of dissipation across the system boundary gives rise to various macroscopic properties of the system, like temperature and pressure that can be statistically estimated. Therefore, it is crucial to define what a system boundary actually is. The concept of boundary is vague from a thermodynamic standpoint. Therefore, the above definitions of a system are inadequate as the Second Law, and the Action principle will readily direct the system to a state of perfect equilibrium with the surroundings (because gradients will eventually disappear). In order to self-contain the definition of a system in terms of structure and interaction, we introduce the concept of 'organization' [46]. The oldest representation of organization can be found in the ancient Greek symbol of *Ouroboros* in the form of a snake devouring its own tail [52, 53]. The symbol represents cyclicity, closure, or simply put, a confinement. In mathematics, the idea of closure is often used to describe algebraic structures, like fields, groups, rings, lattices, vector spaces, *etc*. Using this analogy, organization can be viewed as a causal chain of actions, which eventually *closes onto itself*. Thus, organization not only represents an order, but an order to perform a particular function [12, 13]. We must note at this point that, unlike action, organization is a collective property, which means it is a property observed in a collection of system particles behaving as a coherent whole. In Chemical Organization Theory, the functional aspects of organization and semi-organization are similarly defined. A self-maintaining closed set of molecules is called a semi-organization, whereas a mass-maintaining closed set of molecules is defined as an organization [54, 55]. The property of closure – which is a common thread running across all definitions of organization – assures that the set of molecules contains all the molecules that can be produced by reactions among all those molecules [55].

The property of closure or organization is extremely crucial in the definition of a system. Note that the definition of a closure takes into account the aspects of both structural as well as functional cyclicity. Dissipative systems, like the human body or a cell, can survive and function due to the presence of closures of both the forms [57, 58]. Functional closure is defined as a closed cycle of processes that do not form a physical mediating layer, such as catalytic reactions in an autocatalytic set. In contrast to this, structural closure demands that the interactions create a physically closed topology [58]. Dissipative systems, even though they are open, are resilient to changes in the surrounding media. The presence of feedback loops and response to action[6]mechanisms make these systems to continuously organize (and re-organize) themselves, resulting in the emergence of newer properties and patterns. This property of emerging order (at a global level) out of random interactions (at a local level) is known as self-organization [12, 60]. However, for a system to show self-organizing behaviour, and therefore, 'live', the structural closure should be able to preserve its functionality or the functional closures within the system.

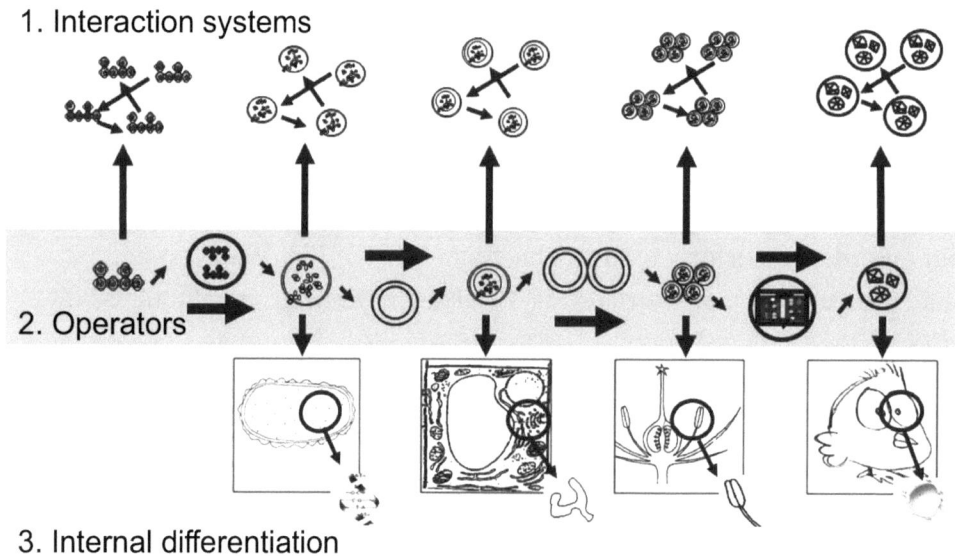

1. Interaction systems

2. Operators

3. Internal differentiation

Fig. (5). Figure represents the three dimensions of complexity: inward (due to internal differentiation), outward (due to formation of more complex structures, unicellulars to multicellulars) and upward (due to first-next-possible closures) [56].

In our discussions, we did not include the concept of growth in complexity in a system, such as unicellulars to multicellulars, or atom to molecules (see Fig. **5**). In order to understand what complexity is from a system-organization perspective, we need to define the smallest entity or a building block that is capable of transcending across the hierarchy of the organizational ladder (see Fig. **6**). The smallest such unit is known as the operator [58, 61]. An operator can be defined as an entity that exhibits first-next-possible (FNP) closure with a closure dimension of atleast three. FNP-closure is the very next *first* possible closure which an existing entity can create in order to increase in the organizational hierarchy. Whereas, closure dimension focuses on the similarity in structural closures, functional closures or a combination of both, such as the presence of only a mediating layer or a hypercycle mediating interface (HMI). Both, senescence and cell death, can be formulated as a cell's gradual loss of structural closure or an inability to maintain its functionality (functional closure) due to the loss of complexity, unconstrained degradation of free energy or reduced resource dominance due to ageing. The distinction between life and death can be directly attributed to the loss of structural closure or to the efficiency of the operator; a loss in structural closure will eventually result in the loss of a cell's functionality [61]. In this section we looked at the energy-entropic perspective of a CAS. We formulated the process of life as an energy dispersing mechanism, and related the process of ageing to the fundamental laws of physics by describing senescence more as a holistic physical process than a purely biological one.

COMPLEXITY: INTERACTIONS

In the previous section, we discussed CAS from the perspectives of energy and entropy. We also extended our arguments to understand the behaviour of cell death and ageing. Towards the end of the earlier section, we tried to visualize the order of complexity through a hierarchy based upon our definition of organization and closure dimensions (see Fig. **6**). In this section, we will focus on interactions between a system's constituent particles and try to quantify the notion of complexity. It is hard to predict the behaviour of a complex system precisely because they are indeterministic, dynamic and nonlinear. In order to understand the evolution and the time-dependent behaviour of these systems, we need various mathematical modelling or simulation methods, like cellular automata, differential equations or agent based modelling techniques. Amongst the existing mathematical methods, network theory in particular has been immensely successful in describing real world systems and processes in recent times. Be it social ties or technological interactions, socio-economic infrastructures or interaction between biological entities, everything can be modelled as a network containing nodes and edges [62 - 71]. One of the very important properties that these network models capture is the way different agents interact with each other in a connected system, which gives rise to non-trivial (emergent) properties in the system. These models also help in identifying the relative importance of the agents within the system, which in turn help to understand the dynamics of the systems' behaviour when subjected to external perturbation [62, 72]. In this section, we incorporate the network modelling (science) concepts into the biological domain to quantify complexity, and to extend our arguments to evaluate ageing from this perspective.

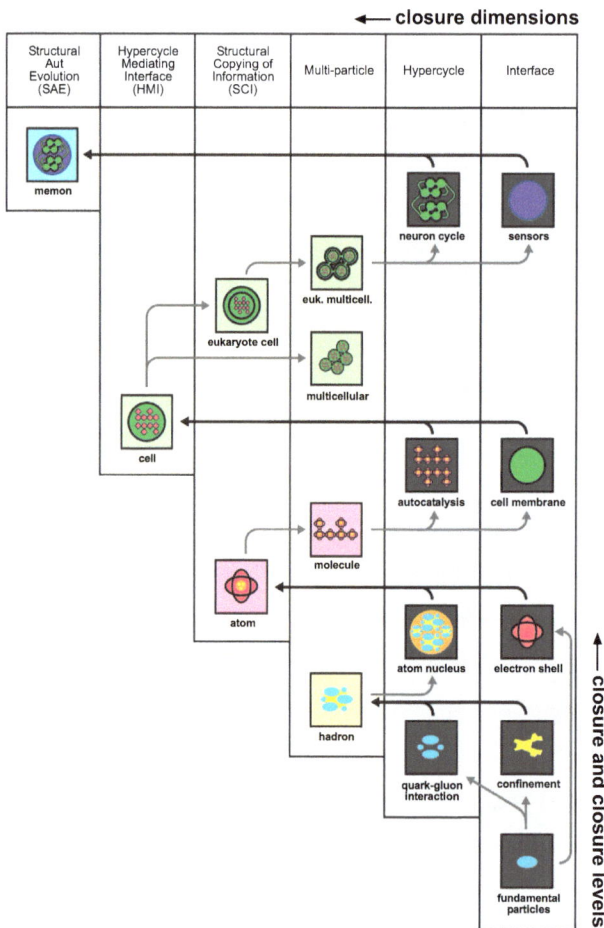

Fig. (6). Figure shows the evolution of operators. The dark line describes the pathway along the subsequent first-next-possible closures. SAE = Structural Auto Evolution, SCI = Structural Copying of Information, HMI = Hypercycle Mediating Interface, CALM = Categorizing And Learning Module [56].

The reason that network science has turned out to be immensely successful these days lies in its simplicity and universality in its application across domains. The underlying concept in network theory is very logical and easy to implement. Any complex system is made up of a large number of parts or agents which interact with each other. In the network science framework, every agent is identified as a node (N), and the interactions between any pair of nodes is considered as a link (L). Therefore, a complex system is modelled as a graph containing nodes and links. Depending upon the type of the problem, graphs can be either unidirectional (directed) or bidirectional (undirected), bipartite or multipartite. In some cases, the links can be associated with weights (weighted graphs), such as the number of papers co-authored by a pair of authors in a collaboration network or the strength of interaction between two proteins in a protein-protein interaction network. In all of these cases the topology of the graph strictly depends upon how the nodes are connected to each other. The degree of a node is the number of connections a node shares with the other nodes in the graph. The pattern of connectivity of the nodes in a network is represented by its degreedistribution function. Based upon the degree-distribution function[7]alone, networks can be classified as random or small-world.

The **small-world** property is of significant interest in understanding the growth of networks and the relative change in the network diameter. It has been observed that in networks exhibiting a small-world phenomenon, the network diameter scales as the logarithm of the network size[8], or $D \sim \log(N)$ [62, 63]. In order for the networks to show this kind of typical scaling relationship, they need network growth mechanisms that fall into either of the two network models: (**i**) the Watts-Strogatz small world model (WS), or (**ii**) the Barabasi-Albert model (BA). Interestingly, real-world networks tend to follow the BA model, which advocates a heterogeneity in the degree-distribution patterns through the networks [62]. While a small subset of the nodes (hubs) share very large number of connections, the majority of the nodes tend to have fewer connections. The presence of *preferential attachment* among the nodes in these networks gives rise to a power-law degree-distribution pattern. This provides us with two interesting insights: (**i**) not all nodes are the *same* in a network; some nodes are more *central* than the others, and (**ii**) the presence of a simple rule like the preferential attachment which shows self-organizing behaviour in real-world network models [62].

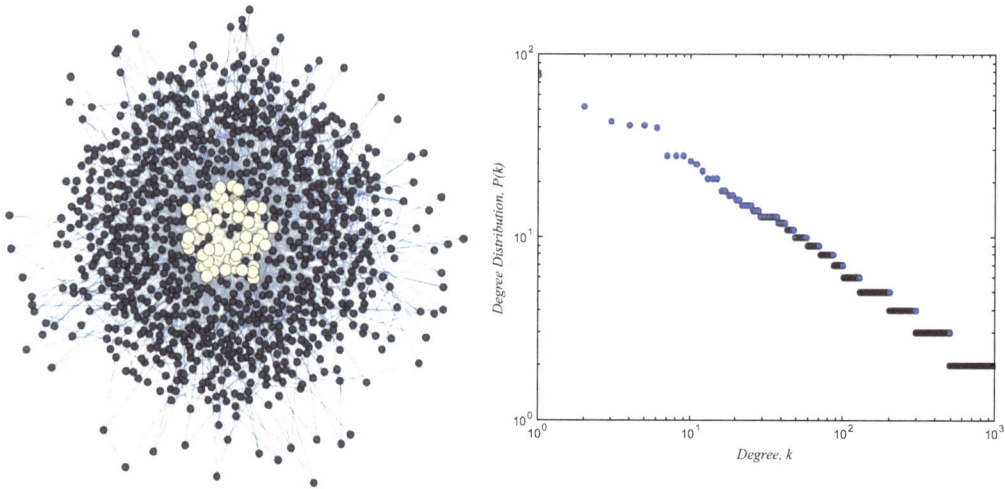

Fig. (7). The left panel shows a scale-free network with $N = 1000$ and $L = 3187$. The node size reflects the node degree, *i.e.*, nodes which have bigger size have higher degrees and vice versa. The color of the nodes partition the network into two distinct communities. In the right panel, we plot the degree-distribution of the network on a double logarithmic scale. The degree-distribution follows a power-law pattern, $P(k) \sim k^{\gamma}$ with $\gamma = 3$. The parameter, γ is calculated from the slope of the degree-distribution plot in the right panel.

We extend our understandings from the network science to biological systems. Many biological structures have been modelled using concepts from networks science, such as the neural architecture of the brain, the protein-protein interaction network in a cell (PPI) or the metabolic networks in a living cell Fig. (**7**). In all of these and many other similar systems, the nodes are either the neurons, the proteins or the enzymes, whereas the interactions are mapped from the actual physical interactions between neurons, protein-protein interactions or from metabolic reaction sets. In all the above systems, the common network properties of relative node importance (or centrality) and self-organizing behaviours have been observed. The inter-connectivity between the agents of a networked system plays a key role. One of the interesting observations in PPI networks is the *centrality-lethality rule*, according to which, in a network, the more central a protein, the more lethal its deletion from the network will be [65] Fig. (**8**. Thus, the 'hub' proteins are relatively more important in the PPI network than the other remaining ones due their larger connectivity. The removal of a hub will turn out to be lethal, as it may result in a disintegrated network with separate unconnected components. On the contrary, the hub proteins tend to attract other proteins, thereby increasing their degree in the process. The process of self-organization, thus operating in the underlying network, gives rise to topologically organized structures; structures which are efficient as the inter-nodal distances considerably shrink as the logarithm of the network size. Presence of such small topological distances and high clustering between the nodes (both the properties are observed in small-world networks) of the same kind make these networks robust.

In neural networks, this kind of structural advantage allows more efficient signal transmission between the neuronal synapses, and high connectivity between the neurons provide stability to random failures [78].

In the previous section, we discussed a complex system from a physical perspective using definitions of action and closures. We observed how the complexity grows due to internal differentiation, addition of FNP-closures and through emergent group behaviour. In this section, we tried to model the complexity associated with these systems into a mathematical framework of graphs. In both the above discussions, the process of self-organization plays a key role, either by adapting to the surrounding perturbations through export of energy and entropy, or through reduced inter-nodal distances due to preferential attachment rules. Both the above phenomena address the system's robustness and resiliency. The question of ageing or gradual decay of the system was attributed to the system's loss of closure or reduced resource dominance. Similarly, the network model predicts the long range evolution of the system based upon the fitness of the nodes to acquire links and the competition between newer nodes to acquire existing links. A decaying, ageing system will tend to have ageing nodes with reduced fitness to attract links. A loss of fitness will result in reduced self-organization, and therefore, reduced efficiency, increased topological inter-nodal distances, reduced resiliency, robustness, and hence, reduced lifespan.

Complexity: Information and Entropy

Twentieth-century theoretical physics came out of the relativistic revolution and the quantum mechanical revolution. It was all about simplicity and continuity (in spite of quantum jumps). Its principal tool was calculus. Its final expression was field theory. Twenty-first-century theoretical physics is coming out of the chaos revolution. It will be about complexity and its principal tool will be the computer. Its final expression remains to be found. Thermodynamics, as a vital part of theoretical physics, will partake in the transformation [79].

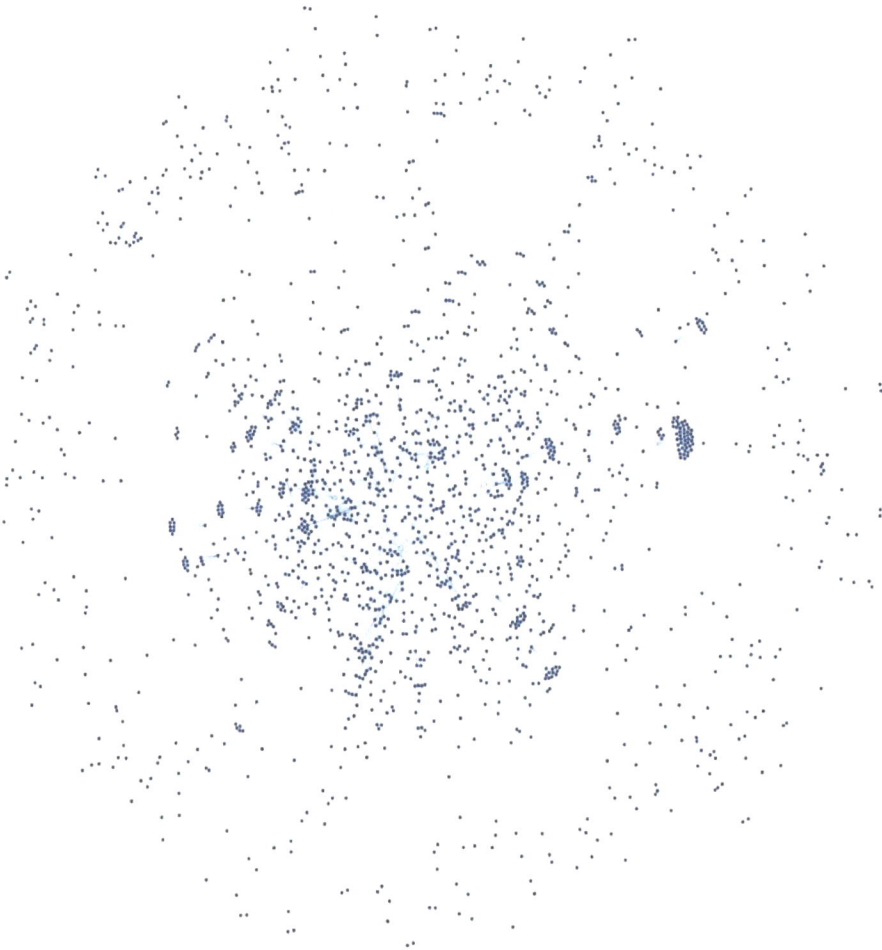

Fig. (8). Figure shows the PPI network structure of the yeast *Saccharomyces cerevisiae*. The nodes ($N = 1870$) represent the proteins and the links ($L = 2240$) the physical interactions between them. The network shows strong assortativity with $r = 0.75$ and a scale-free degree-distribution pattern with an exponential cut-off ($k_c = 20$), $P(k) \sim \exp(-k_c/k)k^{-\gamma}$. Random mutations in the genome of *S. cerevisiae*, modelled by the removal of randomly selected yeast proteins, do not affect the overall topology of the network, whereas when the most connected proteins are computationally eliminated, the network diameter increases rapidly. This simulated tolerance against random mutation is in agreement with results from systematic mutagenesis experiments, which identified a striking capacity of yeast to tolerate the deletion of a substantial number of individual proteins from its proteome [65, 76, 77]. Dataset source: http://www.barabasilab.com/.

In the previous sections, we looked at the structure of a complex adaptive system from two different perspectives (theoretically and computationally). Both the perspectives gave us interesting insights about these systems in terms of metrics of complexity and self-organization. Also, at some level, we could relate one to the other. In this section, we focus entirely on the aspect of disorder or information-theoretic aspect of these systems. The information perspective will look at the concept of entropy from the point of view of chaos and disorder in these systems. Earlier we had remarked about the drawbacks of classical physics, which are reductionism and indeterminism. The definition of entropy as disorder (from a microscopic perspective) takes into account the indeterministic aspects of a complex system, because interactions are stochastic and the presence of multiple parts further renders uncertainty to the system and makes it nonlinear. The concepts of entropy and information are both very interesting in this context and both of them are intimately connected to the study of complex systems [14, 79]. Entropy is supported by the Second Law of Thermodynamics that directs systems towards the state of equilibrium with the surroundings. Thus, any observable differences between the system and the surrounding environment vanish, whereas biological systems generate order from local interactions which is mainly supported by the observations from non-equilibrium thermodynamics.

A question, therefore, arises – what is order? We have spent a great deal of effort to describe life, growth, evolution, ageing and death using energy, entropy and interactions, but we are yet to define or understand the concept of order in complex systems. In this section, we look into this aspect of complex systems in detail and relate the observations to our earlier findings.

The concept of information as a statistical quantity was first used by Shannon to describe the concept of information entropy [80]. In information theory, the (information) entropy measures the expected value of the information contained in the message received, which can be in the form of bits, strings *etc.* Entropy is basically the *uncertainty* in the message received, which is more if the message (say, coin flips) is more random and less if the message received is less random. The information entropy measures the probability of any bit of information stored in a phase space[9]. If the information is uniformly distributed in the phase volume then it is impossible to predict the exact location of the information content. Thus, information has the highest order of disorder in the phase volume, because each point in the phase space is likely to store the information. Using this as our argument, we can define entropy as the lack of information. The resemblance information entropy holds with the statistical formulation of thermodynamic entropy is due to the equation (of the information entropy) that was put forward by Shannon [80]- see appendix below. The resemblance between the two definitions of entropy is not coincidental. We saw earlier (see Fig. **5**) that in the three directions of complexity, the internal differentiation is an important aspect in the complexity growth phenomenon. The information theoretic approach and statistical entropy serve as powerful tools to exploit these aspects of complexity growth, self-organization and scale-invariance.

In the above sections, we tried to determine the essence of complexity, first, from the structural and interactive perspectives (the three dimensions of complexity), and then from the mathematical framework by mapping real-world physical systems into an interaction model of graphs with nodes and links. We were successful to appreciate as well as quantify complexity from both the perspectives, yet a physical expression for complexity was missing in the above formulations. In this section, we develop the physical intuition behind the concept of complexity and relate it with energy-entropy arguments from our previous sections. Complexity of a system can be expressed as a measure of its *thermodynamic depth*, which is mathematically represented as the difference between a system's coarse-grained[10]and fine-grained entropy [82]. Fig. (**9**) the concept of thermodynamic depth relates the entropy of a system to the number of possible historical paths that led to its observed state, with "deep" systems being all those that are complex and "hard to build", and whose final states carry much information about the history leading up to them. The emphasis on how a system evolves with time, its history, identifies the thermodynamic depth as a complementary measure to logical depth. In the real-world we observe biological structures like the brain or artificial structures like the Internet becoming increasingly complex with time. A question, therefore, arises as to how the thermodynamic depth of these systems varies with time, and what is the mechanism behind the increase in complexity of these physically different yet topologically similar structures?

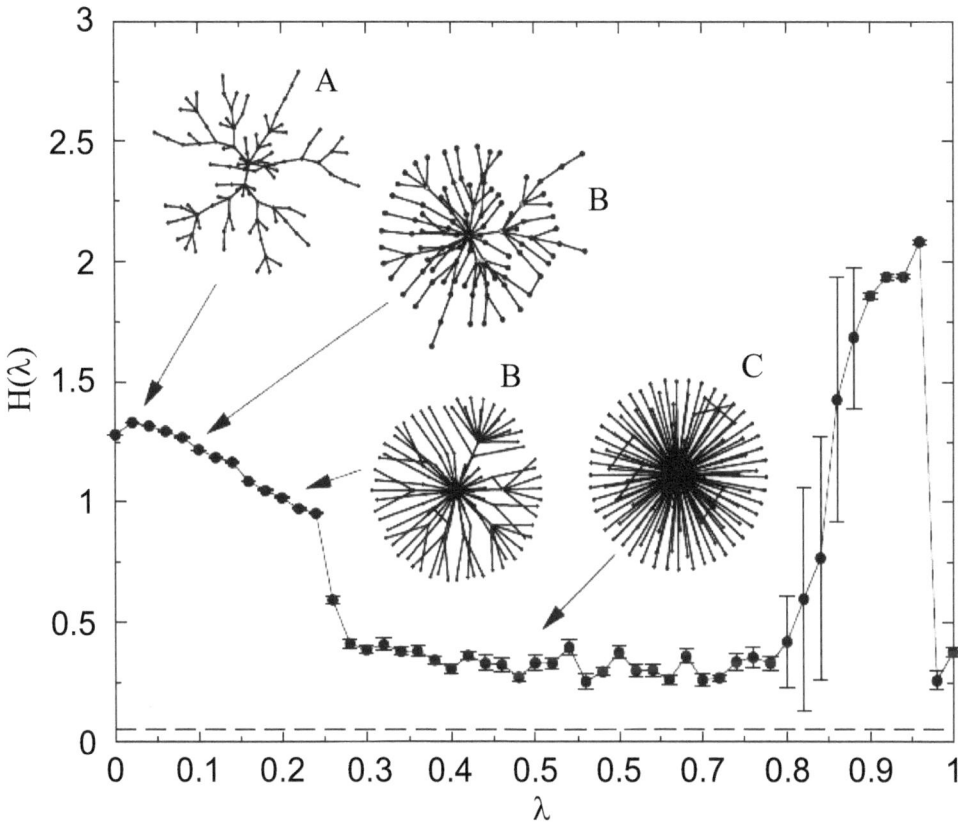

Fig. (9). Figure shows the variation in entropy with different networked structures. The statistical entropy is given as $H(\lambda) = -\int P(k)\log P(k)dk$, where $P(k)$ is the degree-distribution function of the networks. Note that entropy varies with the parameter, λ that controls the network structure based on the following optimization program: $\min E(\lambda) = \lambda l + (1 - \lambda)\rho$, where E is the energy-function, l is the network path-length (diameter), ρ is the network density, and $0 \leq l,\rho,\lambda \leq 1$. The entropy of scale-free networks is lower than equivalent exponential graphs, which can be attributed to self-organization [83].

Greater levels of complexity are achieved in those systems that exhibit a mixture of order and disorder (randomness and regularity), and also have an increased capacity to generate emergent phenomena due to self-organizing processes. Liouville's theorem states that the phase space (volume) should be conserved in time, which means that the entropy (of a phase space or a system) should be conserved [23]. This is a terrible contradiction as our entire discussions are solely based upon the fact that the entropy of a closed system should always increase. The solution to this paradox can be resolved by unifying all the three approaches (in this chapter) to understand complexity. The solution to the above paradox lies in the thermodynamic depth of the system: the 'deeper' a system, the more complex it is. Here, a system's depth implies internal differentiation within the system (three dimensions of complexity). For the internal scales of the system to sustain, the corresponding closures should be reinforcing. Finally, internal differentiation will result into self-similar structures (fractals) or 'pockets' (more microstates) within the system [84]. Each of these pockets will now be more likely to contain the information. Since, the original system (phase space) gets fractalized into self-similar components, the phase volume increases, which increases the corresponding entropy, yet satisfying Liouville's theorem. The presence of more 'pockets' implies more *storage space* for information, more 'depth', increased self-similarity and inter-connectedness, which further implies higher complexity.

DISCUSSION

This chapter aims to present new insights into the biological concept of ageing. In order to develop our ideas about the subject, we look into the fundamental laws of nature. Since every physical, chemical or biological process can be visualized as a transaction of energy and entropy (between the system and its surroundings), we believe that the natural process of ageing and death must also have similar conceptual roots. In order to substantiate our argument with valid reasoning, we first developed the concept of ageing and death in a complex adaptive system. It is well known that the human body and the brain are complex adaptive systems. But the questions that we propose to answer through this chapter are: Why do we age? Why does the robustness of the human body (or, for the matter, all complex adaptive systems) decrease with time? These kinds of general questions are not easy to answer. As the human body is a complex adaptive system, it contains numerous parts (agents/entities) which interact with each other in non-trivial ways at all levels of complexity (cellular, molecular, organelle, neuronal and social) [14, 85]. Therefore, it is necessary to look at this fundamental problem from a systems-perspective by taking into account the role of the various agents, their interactions with each other, their complexity at each level and the corresponding emerging properties (arising due to interactions at each level) due to self-organization.

We tackle this challenge by focusing on the physical laws that govern the natural processes in the universe: the Principle of Least Action and the Second Law of Thermodynamics. Our choice of looking at the biological problems of ageing and rejuvenation from the above physical principles is justified in their infallibility. Like every other thermodynamic system and process, the human body and the purpose of life is to degrade free-energy in the least possible time. It is interesting to note that although the purpose of life is so obvious, yet we see complexity and diversity in life forms around us. To answer this non-trivial question, we needed to give a closer look into the laws of thermodynamics from a system-theoretic perspective. We argued that the presence of closures – structural and functional – act as energy barriers to the otherwise spontaneous dissipation of energy from the system into the surroundings. We looked at the hierarchy of organizational ladder (see Fig. **6**) and observed different levels of complexity, which then motivated us to investigate another fundamental question, *i.e.*, what is (are) the mechanism(s) of complexity-increase in nature? To answer this question, we looked at the three dimensions of complexity (see Fig. **5**). Since a CAS contains multiple parts that interact with each other, we needed an explanation to relate the dimensions of complexity with the mutual interactions of the agents. Also, we further needed to look at the mathematical foundations to understand the underlying process of self-organization in these systems. The answer to the above question also holds the key to the answer of our primary problem that we address in this chapter.

We extended our reasoning to the microscopic level to visualize interactions in these systems. The tool that we used for this purpose is state of the art and something which holds great promises for many future breakthroughs. Using the concepts of network science, we observed that all the nodes (proteins, neurons, agents) do not enjoy the same level of importance. While removal of some nodes may result in a system's 'death', the removal of others might be completely insignificant (centrality-lethality rule). We also observed that real-world networks (interestingly, both biological as well as technological) follow simple rules while acquiring links (resulting into power-laws), thus showing self-organizing behaviours. Power-laws, scale-invariance (self-similarity or fractality) and self-organization are very intimately connected [86, 87]. From systems' perspective, the process of self-organization is a system's spontaneous 'motion' towards a state of *organized* equilibrium through reduction of constraints and dispersal of energy along the least time trajectories [88]. This time-asymmetric evolution of the system converts chaos into order by increasing its thermodynamic depth. The internal differentiation process repeatedly fractalizes the structure of the system by increasing the coarse-grained entropy and reducing the fine-grained entropy. However, all these processes tend to happen within the energy barriers of the structural closure [47, 56, 80]. The loss of closure will result into the system's spontaneous conversion into a state of equilibrium with the surroundings. The presence of attractors in the self-organizing process reduces the system's degree of freedom or its agility, yet increasing its complexity. This gradual loss of agility will cause the system to eventually become fragile, thus making the nodes unfit to acquire newer links. This will increase the topological distances in the network, and will result in reduced efficiency in information transmission across nodes. The above sequence of events is what the process of ageing is comprised of.

In today's technology-laden environment, we are exposed to large quantities of information. This structured way of information has led to increased biological complexity and functionality, yet tackling the problem of ageing is a medical challenge [90, 91]. In spite of our technological and medical breakthroughs, the fate of physical systems and processes are certain: energies will disperse off along the paths of least action, entropy of closed systems will continue to increase, complexity, as well, will continue to increase, and self-organization, as a guiding principle will drive systems towards states to reduced agility. In today's world, where the definitions of disciplines are changing and the boundaries vanishing, we need to look at this problem of increasing lifespans from a holistic framework [92]. One of the exciting new domains which hold great promises is the upcoming field of network medicine, where network science concepts are used to identify, prevent and treat diseases [93, 94]. Therefore, the time is ripe for us to understand all the dimensions of the process of ageing, and fine tune our mathematical and biological tools to challenge this natural process holistically.

ACKNOWLEDGEMENTS

The author wishes to acknowledge the discussions with Marios Kyriazis, Georgi Georgiev, Arto Annila, Gerard Jagers, Quiping Wang, Francis Heylighen and Adrian Bejan at different points in time, which helped in shaping the manuscript in the present form. The author thanks Gerard Jagers for granting him the permission to use the figures from his Ph.D. thesis and Inge Jagers for providing a fine tuned version of the Operator Hierarchy evolution diagram. The author acknowledges the support from the ECCO-GBI group for their excellent hospitality during his stay in Brussels and Gerard and Florentine Jagers for hosting him in The Netherlands. Finally, the support from Indian Institute of Technology Madras is gratefully acknowledged.

NOTES

[1] Muller's ideas appeared in 1922 and were refined again in 1929, long before Schrdinger's work in 1944. Muller also criticized Schrodinger's work in a letter to a journalist stating that Schrodinger's "What is life?" was partly an extension of works published before 1944, and the rest of the book constituted of incorrect speculations [3].

[2] Since biological processes on earth take place at a constant temperature and pressure, Gibb's free energy seems to be a suitable metric instead of Helmholtz's free energy (a similar metric, which is calculated under the conditions of constant volume and temperature). The change in Gibb's free energy is mathematically given as, $\delta G = \delta H - T \delta S$, with H being enthalpy, T temperature and S entropy [19].

[3] Out of all the possible interpretations, Hamilton's and Maupertuis' formulations are particularly interesting. Where Hamilton's formulation explicitly fixes the end points, Maupertuis' formulation is more relaxed. Also, note that Euler's and Maupertuis' formulations are mathematically equivalent.

[4] Accessed from the homepage of Prof. Arto Annila, University of Helsinki, Finland: http://www.helsinki.fi/~aannila/arto/

[5] A chemical reaction, say A + B → C proceeds in the forward direction spontaneously if the product obtained (C) has a lower free-energy than the reactants (A,B). The above process seems obvious yet happens along a non-trivial trajectory. When the reactants combine, their total freeenergy increases as compared to their initial states, which is in conflict with the Gibb's equation and the Second Law. This energy barrier or a threshold, which the reactants need to cross, is known as the *activation energy*. Once this barrier is crossed the reaction occurs spontaneously [19].

[6] By response to action, we mean an open system's response to the surrounding perturbation. The surrounding media is also made up of quanta or units of action. The exchange of quanta between the system and the surrounding will cause the system's symmetry to spontaneously break, thus, generating new organizations in terms of functional closures [59].

[7] The degree-distribution function, $P(k)$ can be used to directly categorize networks. Networks with degree-distribution patterns following a power-law, $P(k) \sim k^{-\gamma}$ are scale-free and show scale invariance, networks with exponential degree-distribution patterns, $P(k) \sim \exp(-\lambda k)$ tend to be random [62, 73]. Also, note that $^{R}P(k)dk = 1$.

[8] Based on the scaling parameter, γ, networks have been found to be ultra-small. For $2 < \gamma < 3$, it has been observed that $D \sim \mathrm{loglog}(N)$ [74, 75].

[9] Mathematically, a phase space of a dynamical system is a space in which all possible states of a system are represented. Each possible state of the system corresponds to a unique point in the phase space. Thus, the phase space captures the trajectory of the system's evolution with time (or other variables on which the system's dynamics depend upon). A physical system can also be considered as a phase space (volume) where each of its constituting elements are the microstates, with each of them being identified by a unique pair of their momentum and position vector. Thus, all possible configurations of the system are called statistical ensembles. The above definition extends to the concept of networks,

where a random graph with a particular degree-sequence is identified as an instance of the statistical ensemble of the possible configurations of the graph [23, 47, 81].

[10] The coarse-grained entropy is the statistical entropy of the system, whereas the fine-grained entropy is the Boltzmann's constant multiplied by the information entropy. The thermodynamic depth is represented as $d\rho$ = *Sthermo* − *kBSShannon.*

APPENDIX

The (statistical) thermodynamics entropy was given by Boltzmann in terms of microstates, as S = −kBΣpi lnpi where pi, represents the microstates taken from an equilibrium ensemble and kB, the Boltzmann's constant. The information entropy formulation by Shannon was somewhat similar to the Boltzmann's formulation of statistical entropy, S = −Σpi lnpi where pi, represents the probability of choosing a message [19, 80]. Since information entropy is the uncertainty involved in the information content, the metric, information can be defined as the reverse of entropy, *i.e.*, I = 1−S.

REFERENCES

[1]   Erwin Schrödinger, Lewin . What is life?. Cambridge University Press 1944.

[2]   Derry J. Book Review: What is Life?. 2004.

[3]   Hyman T, Brangwynne C. In retrospect: the origin of life. Nature 2012; 491(7425): 524-5.

[4]   Schwartz J. In Pursuit of the Gene. Harvard University Press 2009.

[5]   Dronamraju KR. Erwin Schrödinger and the origins of molecular biology. Genetics 1999; 153(3): 1071-6.
        [PMID: 10545442]

[6]   Planck M. Treatise on Thermodynamics. Dover 1945.

[7]   Gibbs JW. The collected works of J Willard Gibbs. Yale University Press 1957; Vol. 1.

[8]   Cengel YA. Introduction to thermodynamics and heat transfer + EES software. New York: McGraw Hill Higher Education Press 2007.

[9]   Clausius R. Uber eine ver¨anderte form des zweiten hauptsatzes der mecha-¨ nischen wa¨rmetheorie. Ann Phys 1854; 169(12): 481-506.
        [http://dx.doi.org/10.1002/andp.18541691202]

[10]   Gibbs JW, Bumstead HA, *et al.* Thermodynamics. Longmans, Green and Company 1906; 1.

[11]   Elliott H Lieb and Jakob Yngvason. The physics and mathematics of the second law of thermodynamics. Phys Rep 1999; 310(1): 1-96.
        [http://dx.doi.org/10.1016/S0370-1573(98)00082-9]

[12]   Heylighen Francis, *et al.* The science of self-organization and adaptivity. The encyclopedia of life support systems 2001; 5(3): 253-80.

[13]   Heylighen F, Joslyn C. Cybernetics and second order cybernetics. Encyclopedia of physical science & technology 2001; 4: 155-70.

[14]   Bar-Yam Y. Dynamics of complex systems. MA: Addison-Wesley Reading 1997; 213.

[15]   Fath Brian D. How system thinking approaches and the notion of energy metabolism of urban socioeconomic sectors can inform energy conservation policies 2015.

[16]   Dorigo M, Maniezzo V, Colorni A. Ant system: optimization by a colony of cooperating agents. Systems, Man, and Cybernetics, Part B: Cybernetics, IEEE Transactions on 1996; 26(1): 29-41.

[17]   Alexander V. Getling Rayleigh-B'enard convection: structures and dynamics. World Scientific 1998; 11.

[18]   Boltzmann L. The second law of thermodynamics. Theoretical physics and philosophical problems. Springer 1974; pp. 13-32.
        [http://dx.doi.org/10.1007/978-94-010-2091-6_2]

[19]   Atkins P, De Paula J. Elements of physical chemistry. Oxford University Press 2013.

[20]   Lehninger Albert L, Nelson David L, Cox MM. Bioenergetics and metabolism. Principles of Biochemistry 1993; 2

[21]   Kozliak E, Frank L. Lambert. Residual entropy, the third law and latent heat. Entropy (Basel) 2008; 10(3): 274-84.
        [http://dx.doi.org/10.3390/e10030274]

[22]   Frank L. Lambert. The conceptual meaning of thermodynamic entropy in the 21st century. Int Res J Pure Appl Chem 2011; 1(3): 65-8.
        [http://dx.doi.org/10.9734/IRJPAC/2011/679]

[23]   Goldstein H. Classical mechanics. Pearson Education India 1957.

[24]   Richard P. Feynman, Robert B Leighton, and Matthew Sands The Feynman Lectures on Physics, Desktop Edition. Basic Books 2013; Vol. I.

[25]   Moreau de Maupertuis Pierre-Louis. Essais de cosmologie. 1751.

[26]   Euler L. Methodus inveniendi lineas curvas maximi minimive proprietate gaudentes sive solutio problematis isoperimetrici latissimo sensu accepti. Springer Science & Business Media 1952; Vol. 1.

[27]   Lagrange JL. M'ecanique analytique. Mallet-Bachelier 1853; Vol. 1.

[28]   Moreau de Maupertuis Pierre-Louis. Accord de diff'erentes loix de la nature qui avaient jusqu'ici paru incompatibles. 1744.

[29]   Moreau de Maupertuis P-L. Les loix du mouvement et du repos d'eduites dun principe m'etaphysique. Histoire de lAcad'emie Royale des Sciences et des Belles Lettres. 1746; pp. 267-94.

[30]   Feynman RP. Albert R Hibbs, and Daniel F Styer Quantum mechanics and path integrals: Emended edition. Dover Publications 2005.

[31]   John A Wheeler and Edwin F Taylor. Spacetime physics. WH Freeman 1966.

[32]   Ville RI Kaila and Arto Annila. Natural selection for least action.

[33]   Grönholm T, Annila A. Natural distribution. Math Biosci 2007; 210(2): 659-67.
        [http://dx.doi.org/10.1016/j.mbs.2007.07.004] [PMID: 17822723]

[34]   Natural emergence. Complexity 2012; 17(5): 44-7.
        [http://dx.doi.org/10.1002/cplx.21388]

[35]   Georgiev G, Georgiev I. The least action and the metric of an organized system. Open Syst Inf Dyn 2002; 9(04): 371-80.
        [http://dx.doi.org/10.1023/A:1021858318296]

[36]   Georgiev GY. A quantitative measure, mechanism and attractor for self-organization in networked complex systems. Self-Organizing Systems. Springer 2012; pp. 90-5.
        [http://dx.doi.org/10.1007/978-3-642-28583-7_9]

[37]   Georgiev GY, Henry K, Bates T, *et al.* Mechanism of organization increase in complex systems. Complexity 2014.

[38]   Georgiev G. Spontaneous symmetry breaking of action in complex systems. Bull Am Phys Soc 2014; 59(1)

[39]   Lucia U. Probability, ergodicity, irreversibility and dynamical systems. Proceedings of the Royal Society A: Mathematical, Physical and Engineering Science. 464(2093): 1089-104.
        [http://dx.doi.org/10.1098/rspa.2007.0304]

[40]   Lucia Umberto. Maximum or minimum entropy generation for open systems? Physica A: Statistical Mechanics and its Applications 2012; 391(12): 3392-8.

[41]   Lucia Umberto. Mathematical consequences of gyarmatis principle in rational thermodynamics. Il Nuovo Cimento B Series 11 1995; 110(10): 1227-35.

[42]   Qiuping A. Wang. Incomplete statistics: nonextensive generalizations of statistical mechanics. Chaos Solitons Fractals 2001; 12(8): 1431-7.
        [http://dx.doi.org/10.1016/S0960-0779(00)00113-2]

[43]   Chatterjee A. Certain interesting properties of action and its application towards achieving greater organization in complex systems. arXiv preprint arXiv:1111 2011; 3186

[44]   Chatterjee A. Action, an extensive property of self–organizing systems. International Journal of Basic and Applied Sciences 2012; 1(4): 584-93.
        [http://dx.doi.org/10.14419/ijbas.v1i4.419]

[45]   Chatterjee A. Principle of least action and theory of cyclic evolution. arXiv preprint arXiv:1111 2011; 5374

[46]   Chatterjee A. Principle of least action and convergence of systems towards state of closure. International Journal of Physical Research 2013; 1(1): 21-7.
        [http://dx.doi.org/10.14419/ijpr.v1i1.779]

[47]   Chatterjee A. On the thermodynamics of action and organization in a system. Complexity 2015. under review
[http://dx.doi.org/10.1002/cplx.21744]

[48]   Sharma V, Annila A. Natural process--natural selection. Biophys Chem 2007; 127(1-2): 123-8.
[http://dx.doi.org/10.1016/j.bpc.2007.01.005] [PMID: 17289252]

[49]   Chatterjee A. Intrinsic limitations of the human mind. International Journal of Basic and Applied Sciences 2012; 1(4): 578-83.
[http://dx.doi.org/10.14419/ijbas.v1i4.418]

[50]   Chatterjee A. Is the statement of murphy's law valid? Complexity 2015.
[http://dx.doi.org/10.1002/cplx.21697]

[51]   Chatterjee A, Georgiev G. Physical foundations of self-organizing systems. APS March Meeting Abstracts. 1: 1264P.

[52]   Khlopov M. Fundamental particle structure in the cosmological dark matter. Int J Mod Phys A 2013; 28(29): 1330042.
[http://dx.doi.org/10.1142/S0217751X13300421]

[53]   De Chardin PT, Wall B, *et al.* The phenomenon of man. Row New York, NY, USA: Harper 1965; 383.

[54]   Heylighen Francis, Beigi Shima, Veloz Tomas. Chemical organization theory as a universal modeling framework for interaction, self-organization, and autopoiesis 2015.

[55]   Dittrich P, di Fenizio PS. Chemical organisation theory. Bull Math Biol 2007; 69(4): 1199-231.
[http://dx.doi.org/10.1007/s11538-006-9130-8] [PMID: 17415616]

[56]   GAJM Jagers op Akkerhuis. The operator hierarchy: a chain of closures linking matter, life and artificial intelligence. PhD Thesis.

[57]   Gerard AJM Jagers op Akkerhuis. The pursuit of complexity: the utility of biodiversity from an evolutionary perspective. 2012.

[58]   Jagers op Akkerhuis GA. Analysing hierarchy in the organization of biological and physical systems. Biol Rev Camb Philos Soc 2008; 83(1): 1-12.
[PMID: 18211280]

[59]   Noether Emmy. Invariante variationsprobleme. Nachrichten von der Gesellschaft der Wissenschaften zu G¨ottingen, mathematisch-physikalische Klasse 1918; 1918: 235-57.

[60]   Nicolis G, Prigogine I, *et al.* Self-organization in nonequilibrium systems. New York: Wiley 1977; Vol. 191977.

[61]   Jagers op Akkerhuis. Gajm . Towards a hierarchical definition of life, the organism, and death. Foundations of Science 2010a; 15(3): 245-62.

[62]   Albert R, Barabasi AL. Statistical mechanics of complex networks Rev Mod Phys 2002; 74(1): 47.
[http://dx.doi.org/10.1103/RevModPhys.74.47]

[63]   Strogatz SH. Exploring complex networks. Nature 2001; 410(6825): 268-76.
[http://dx.doi.org/10.1038/35065725] [PMID: 11258382]

[64]   Bork P, Jensen LJ, von Mering C, Ramani AK, Lee I, Marcotte EM. Protein interaction networks from

yeast to human. Curr Opin Struct Biol 2004; 14(3): 292-9.
[http://dx.doi.org/10.1016/j.sbi.2004.05.003] [PMID: 15193308]

[65]   Jeong H, Mason SP, Barabási AL, Oltvai ZN. Lethality and centrality in protein networks. Nature 2001; 411(6833): 41-2.
[http://dx.doi.org/10.1038/35075138] [PMID: 11333967]

[66]   Stam CJ. Modern network science of neurological disorders. Nat Rev Neurosci 2014; 15(10): 683-95.
[http://dx.doi.org/10.1038/nrn3801] [PMID: 25186238]

[67]   Barabási A-L, Oltvai ZN. Network biology: understanding the cell's functional organization. Nat Rev Genet 2004; 5(2): 101-13.
[http://dx.doi.org/10.1038/nrg1272] [PMID: 14735121]

[68]   Pastor-Satorras R, Vespignani A. Epidemic spreading in scale-free networks. Phys Rev Lett 2001; 86(14): 3200-3.
[http://dx.doi.org/10.1103/PhysRevLett.86.3200] [PMID: 11290142]

[69]   Bollen J, Mao H, Zeng X. Twitter mood predicts the stock market. J Comput Sci 2011; 2(1): 1-8.
[http://dx.doi.org/10.1016/j.jocs.2010.12.007]

[70]   Chatterjee A. Scaling laws in chennai bus network. arXiv preprint arXiv:150803504 2015.

[71]   Jiang J, Calvao M, Magalhases A, *et al*. Study of the urban road networks of le mans. arXiv preprint arXiv:10020151

[72]   Mark EJ. Newman. The structure and function of complex networks. SIAM Rev 2003; 45(2): 167-256.
[http://dx.doi.org/10.1137/S003614450342480]

[73]   Deng Weibing, Li Wei, Cai Xu, Qiuping A. The exponential degree distribution in complex networks: Non-equilibrium network theory, numerical simulation and empirical data. Physica A: Statistical Mechanics and its Applications 2011; 390(8): 1481-5.

[74]   Chung F, Lu L. The average distances in random graphs with given expected degrees. Proc Natl Acad Sci USA 2002; 99(25): 15879-82.
[http://dx.doi.org/10.1073/pnas.252631999] [PMID: 12466502]

[75]   Cohen R, Havlin S. Scale-free networks are ultrasmall. Phys Rev Lett 2003; 90(5): 058701.
[http://dx.doi.org/10.1103/PhysRevLett.90.058701] [PMID: 12633404]

[76]   Winzeler Elizabeth A, Shoemaker Daniel D, Astromoff Anna , *et al*. Functional characterization of the s. cerevisiae genome by gene deletion and parallel analysis. Science 1999; 285(5429): 901-6.

[77]   Giaever Guri, Chu Angela M , Ni Li, *et al*. Functional profiling of the saccharomyces cerevisiae genome. Nature 2002; 418(6896): 387-91.

[78]   Eguíluz VM, Chialvo DR, Cecchi GA, Baliki M, Apkarian AV. Scale-free brain functional networks. Phys Rev Lett 2005; 94(1): 018102.
[http://dx.doi.org/10.1103/PhysRevLett.94.018102] [PMID: 15698136]

[79]   Baranger M. Chaos, complexity, and entropy. Cambridge: New England Complex Systems Institute 2000.

[80]   Claude E. Shannon. Prediction and entropy of printed english. Bell Syst Tech J 1951; 30(1): 50-64.
[http://dx.doi.org/10.1002/j.1538-7305.1951.tb01366.x]

[81]   Bianconi G. The entropy of randomized network ensembles. EPL 2008; 81(2): 28005. [Europhysics Letters].
[http://dx.doi.org/10.1209/0295-5075/81/28005]

[82]   Lloyd S, Pagels H. Complexity as thermodynamic depth. Ann Phys 1988; 188(1): 186-213.
[http://dx.doi.org/10.1016/0003-4916(88)90094-2]

[83]   Ferrer i Cancho Ramon. Optimization in complex networks. Statistical mechanics of complex networks. Springer 2003; pp. 114-26.

[84]   Hidalgo C. Why Information Grows: The Evolution of Order, from Atoms to Economies. Basic Books 2015.

[85]   Heylighen F. Cybernetic principles of aging and rejuvenation: the buffering- challenging strategy for life extension. Curr Aging Sci 2014; 7(1): 60-75.
[http://dx.doi.org/10.2174/1874609807666140521095925] [PMID: 24852018]

[86]   Bak P, Tang C, Wiesenfeld K. Self-organized criticality: An explanation of the 1/f noise. Phys Rev Lett 1987; 59(4): 381-4.
[http://dx.doi.org/10.1103/PhysRevLett.59.381] [PMID: 10035754]

[87]   West GB, Brown JH, Enquist BJ. A general model for the origin of allometric scaling laws in biology. Science 1997; 276(5309): 122-6.
[http://dx.doi.org/10.1126/science.276.5309.122] [PMID: 9082983]

[88]   Ross Ashby W. Requisite variety and its implications for the control of complex systems. Cybernetica 1958; 1: 83-99.

[89]   Pascal R. Kinetic barriers and the self-organization of life. Isr J Chem 2015.
[http://dx.doi.org/10.1002/ijch.201400193]

[90]   Kyriazis M. Technological integration and hyperconnectivity: Tools for promoting extreme human lifespans. Complexity 2014.

[91]   Kyriazis M. Reversal of informational entropy and the acquisition of germ-like immortality by somatic cells. Curr Aging Sci 2014; 7(1): 9-16.
[http://dx.doi.org/10.2174/1874609807666140521101102] [PMID: 24852017]

[92]   Sinatra Roberta, Deville Pierre, Szell Michael, Wang Dashun. Nat Phys 2015; 11(10): 791-6.
[http://dx.doi.org/10.1038/nphys3494]

[93]   Barabási AL. Network medicine--from obesity to the "diseasome". N Engl J Med 2007; 357(4): 404-7.
[http://dx.doi.org/10.1056/NEJMe078114] [PMID: 17652657]

[94]   Barabási AL, Gulbahce N, Loscalzo J. Network medicine: a network-based approach to human disease. Nat Rev Genet 2011; 12(1): 56-68.
[http://dx.doi.org/10.1038/nrg2918] [PMID: 21164525]

CHAPTER 8

Engagement with a Technological Environment for Ongoing Homoeostasis Maintenance

Abstract: Emerging empirical and theoretical thinking about human ageing places considerable value upon the role of the environment as a major factor which can promote prolonged healthy longevity. Our contemporary, 'information-rich' environment is taken to mean not merely the actual physical surroundings of a person but it is also considered in a more abstract sense, to denote cultural, societal and technological influences. This modern environment is far from being static or stable. In fact, it is continually changing in an exponential manner, necessitating constant adaptive responses on behalf of our developmental and evolutionary mechanisms. In the previous chapters we have presented our views about these adaptive responses and mechanisms. Here, I will describe in some more detail these and related mechanisms of how a continual, balanced and meaningful exposure to a stimulating environment, including exposure to 'information-that-requires-action' (but NOT trivial information), has direct or indirect repercussions on several factors (mostly epigenetic) which may then act to prolong healthy longevity. Information gained from our environment acts as a hormetic stimulus which up-regulates biological responses and feedback loops, eventually leading to improved repair of age-related damage. The consequence of this up-regulated information-processing systems may influence resource allocation and redress the imbalance between somatic cell *versus* germ-line cell repairs. This can eventually have evolutionary consequences resulting in the drastic reduction of age-related disease and degeneration.

Keywords: Ambient intelligence, Cognition, Complexity, Germ line repair, Human evolution, Information, Phase transition, Physical exercise, Smart cities, Techno-culture.

INTRODUCTION

The discussion here builds upon elements examined in previous chapters, and

Marios Kyriazis

attempts to elaborate on some more details and mechanisms involved. I will revisit several of these elements and present arguments that support my general hypothesis: that **a worldview based upon purposeful and focused integration with technology which hormetically challenges our cognition, may initiate a shift in evolutionary priorities, resulting in a virtually total elimination of age-related dysfunction**. Here, I must make clear that, in my view, evolution is not a totally random process, as it generally has a tendency towards survival. I define the term 'evolution' as 'the adaptation to changes in the environment so that survival continues'. Therefore, life's 'priorities' are geared towards ensuring survival by any means (such as reproduction for instance). It may be possible to modify the thrust of evolution, and shift the emphasis from the survival of the species to the survival of the individual. Although one of the most obvious possible mechanisms of this appears to be the soma to germ line reassignment of repair resources, there could be other mechanisms, perhaps equally as important. However, in this book I am not going to speculate any further about other mechanisms, and will concentrate on the somatic *vs* germ line argument.

I argued elsewhere that ageing is accompanied by loss of information and complexity *i.e.* increased entropy over time [1] (Box **1**). This loss is rooted in the suboptimal conditions caused by the uneven distribution of resources favouring the survival of the germ line *versus* somatic repair. The rate of somatic repair tends to become progressively compromised as a function of age, resulting in accumulation of damaged biological material that reduces organisation and functionality [2] *i.e.* reduced information content, reduced complexity and thus reduced survival.

If it is assumed that the above concepts are, on the whole, valid, then one may ask the question: Can we intervene in any way in order to accelerate or modify this process? In this book we have proposed and discussed interventions and environmental challenges which increase the information load of the individual and may result in up-regulation of essential repair processes, resulting in clinical benefits. At this point I must re-iterate that a 'Challenge' is a situation that potentially carries biological value for an organism, so that the organism is inclined to act. A challenge provokes action because it represents a situation in which not acting will lead to an overall lower fitness than acting.

Box 1. Three essential elements (or assumptions) need to be considered in the case of human ageing:

1. During human evolution, there is a general tendency to attain higher levels of functional complexity [3],
2. The process of ageing is accompanied by a general loss of this functional complexity [4], and
3. By increasing complexity (either artificially, through information or challenges, or through any other means) age-related dysfunction is minimised [1, 5].

Environmental Enrichment: Speculations, Inferences and the Phase Transition

While the health benefits of Environmental Enrichment (EE) have been well studied, it is not clear what impact an enriched environment has on prolonging healthy lifespan, and certainly it is not clear if EE can prolong human lifespan well beyond the current maximum limit (approximately 115-120 years). I have already discussed some initial facts concerning these mechanisms in the previous chapters. Here, an attempt will be made to further clarify some possible mechanisms involved in this respect, particularly with regards to examining whether EE (and thus information) can have any effects on reducing age-related dysfunction and thus prolong human lifespan beyond the current maximum limits.

In order to place the discussion within a wider framework, I will revisit the general concept of phase transition. A phase transition is a profound structural re-organisation of a system, when there is a large change in resource availability for maintenance [6]. Sudden, rapid-acceleration phase transitions are associated with emergence, mutations, breakthroughs and development of autocatalytic systems which disrupt the status quo and result in new situations which may improve fitness and survival [7]. The study of such threshold dynamics gives interesting insights into plausible mechanisms for artificial manipulation of the process. In order for a phase transition to take place it is necessary to have increased complexity, increased exposure to new challenges and a reduction of available resources [8]. In the case under consideration here, the continuous exposure to new actionable information, and the unrelenting cognitive energy demands associated with it, may lead to a critical point where the value of selective pressure induces a phase transition in certain basic biological parameters [9]. Exposure to information may include, for example, environmental enrichment, as discussed throughout this book. A specific case is that of a hormetic approach which is directed at cognition and encompasses enrichment aided by technological inputs such as digital assistants (mobile internet, for example) in a context of societal and cultural advancement, in effect complementing the Global Brain concept. I have suggested that the continual informational cognitive input may have a positive effect on epigenetic DNA changes which could then act in a way to up-regulate certain ageing-modifying factors. This may be augmented not only by environmental enrichment, but also through cultural and societal developments, and increased use of technology, such as ambient intelligence, smart cities, nanotechnology and synthetic biology.

When new information arrives at the sensory organ, it must be processed and thus the process will inevitably demand energy from any available resources [10, 11]. We may now hypothesize that, if the amount of meaningful and actionable information is sufficiently large, and sufficiently sustained, then there will come a point when the ratio between the resources available to the germ line and resources available to the soma (for information processing AND subsequent somatic cell repair) will be forcibly shifted in favour of the latter. In other words, the process of assimilating new information may initiate a sequestration sequence that diverts resources from germ cells to somatic cells. It is important to emphasize that this is taken in the evolutionary context of increased fitness within a technological, information-laden niche. Thus, information processing carries an adaptive value within this specific niche. It 'nudges' the system to behave in a way we desire. The cybernetic concept of 'nudging' means an external, intentional and controlled intervention that shapes the environment in such a way that it promotes desirable actions.

While it is true that the energy used during signalling constrains the flow of information both within and between neurons, a sustained input of information makes such energy demands on the system that may eventually cause a transition to a new state whereby the energy available to the cell will be (in some way) proportionate to the amount of information which the cell is required to process [12]. This energy may thus be used not only for information processing but, necessarily, also for somatic cell repair in order to make information processing possible. It is clearly unlikely that a high level of efficient information processing takes place in a damaged cell. The cell needs to be repaired (this includes age-related damage repair) in order to be in a position to process the information. Indeed, studies have estimated that 50-80% of energy consumption is allocated to signalling, and the remainder is allocated to cell maintenance and repairs [13].

Here a balance must be found between the energy costs of information processing and the contribution to overall fitness (good function within an environment) made by this information. If the fitness of the organism increases as a result of the new information, then it means that the cost of processing that information was worth it. This is essentially the 'consequence-capture' concept [14] which considers how a system is able to capture benefits associated with a particular action. If the benefits to the system are greater than the energetic cost needed to perform the action, then the action would have been worth performing. This will have a secondary effect on other associated processes such as metabolic, immune, vascular and respiratory mechanisms, all of which must also be repaired. It must be reiterated that this discussion refers specifically to the evolutionary fitness of modern humans who live in an increasingly technological niche, where survival depends on adapting successfully to, ultimately, relentless cognitive information processing.

An enriched environment does not only have positive effects on the brain, but it was also shown to have an impact on other tissues, organs and processes such as:

- vision [15]
- the retina [16]
- immunity [17]
- tissue repair [18]
- inflammatory response [19] and
- other physical parameters [20], thus having a holistic effect on the organism.

Together these show that external stimulation does not only 'rejuvenate' the brain but also other non-neural structures and mechanisms which may have been affected by age-related degeneration. This makes perfect sense because it supports the view discussed earlier about the fact that the neuron cannot function fully unless all other supporting somatic cells and tissues are also healthy and functional (chapter 5, Box 1).

The Future: Cognitive Instead of Physical?

So, now we move into a more focused discussion regarding information and environmental enrichment, a discussion which places more emphasis on cognition rather than on physical activities.

I have already hinted in previous chapters that, as our society becomes increasingly technological, some sections of humanity may need to redefine their health priorities, shifting from an essentially physical to an essentially cognitive worldview. An increased amount of cognitive activities may have physical effects not just on the brain but also on other organs and tissues. While it is known that physical exercise can improve cognitive abilities [21], little is known about the effects of increased cognitive actions on the physical health of the body. I argue that it may be possible to use cognitive challenges (and not physical exercise) in order to achieve certain physical benefits which are needed for the day to day optimal functioning of the body. Some research suggests that this may be possible and clinically relevant. For instance, it was shown that mental imagery improves muscle strength without the need to physically exercise [22]. This may happen through activation of neurological mechanisms, most likely at the cortical level. Regular activation of these cortical regions *via* imagery attenuates muscular weakness. The same argument may be true with respect to other factors. We know that physical exercise up-regulates BDNF action [23] which is what cognitive stimulation also does. We also know that while physical exercise improves immunity [24], an enriched cognitive experience can also result in an improved immunity [25]. Physical exercise improves heart rate variability (a measure of health status), while exposure to enriched and cognitive environments also positively regulates heart rate variability [26]. Whilst these examples are interesting in themselves, it is necessary to study many more examples in order to make a rigorous recommendation in clinical settings. Nevertheless, the prospect exists, that it may be possible to achieve a physical health benefit not only exclusively through a physical action but also through a cognitive one.

It has been shown that cognitive exercise such as speed of information processing can improve physical health parameters in older people, such as vitality, physical functioning and bodily pain, as well as social and emotional functioning [27]. Advanced cognitive training not only improves physical functioning but its effects were found to persist after a ten year period [28]. A significant number of older people who followed a programme of **cognitive** exercises rated their **physical** health and quality of life as excellent or very good [29].

The argument made here is that we may be able to experience physical benefits through physical stimulation or through cognitive stimulation, or both, but not necessarily only through physical stimulation. If this argument proves to be applicable extensively, then it would lend support to the general concept of the health value of cognitive challenges, augmented through living within an increasingly technological society. Here I will repeat my comment taken from chapter 6:

If during our evolution, say, we had to use our brain 20% and perform physical activities 80% of the available time, this ratio would now be inappropriate in a society where there is more emphasis on cognition. As a general guidance and based on the broad concept of the Pareto distribution which is a power law probability distribution (https://en.wikipedia.org/wiki/Pareto_distribution), we may be able to adapt more successfully to a modern environment if we reverse this ratio: use our brain 80% and be physical 20% of our available time. This will still provide a suitable mix of physical and cognitive abilities, at least for a section of humanity. I am not arguing for a total replacement of exercise, but for a general propensity to exercise less and think more. This 80:20 ratio is more appropriate in a word where cognition increasingly carries more value.

With regards to energetics, it was shown that an average event of active cognitive stimulation carries an energetic cost of 27μmol ATP/g/min which is within the same order of magnitude as that required by the muscle during physical exercise: 33μmol ATP/g/min [30]. Studies in chess players who undergo increased cognitive effort show that there are significant physical effects on the body [31] such as increased in heart rate (similar magnitude as light exercise) and a combination of carbohydrate/lipid- dependent energy expenditure (as physiologically seen during light exercise). An old study by Leedy and DuBeck [32] showed that during a chess tournament the breathing rate, muscle contractions and systolic blood pressure of the participants increased considerably, within the same magnitude as that seen in athletes during physical activities.

In any case, there are trade-offs to be considered between physical and cognitive effort. It may be argued that the more emphasis we place on physical effort, the less likely it is that we will also expand our cognition (and vice versa). For instance, Kurzban *et al.* [33] have argued that the effort of performing a task depends on a trade-off between costs and benefits associated with that particular task. It is not only a matter of a physical resource being depleted over time with the continuation of the task, but other factors are implicated, such a mental representation of the cost-benefit, and an assessment of the 'worth' of each task. The system will have to assess the cost and somehow 'decide' if the effort associated with the task is worth it. Therefore, before choosing to perform a task, our executive functions will assess if it is best to perform a physical action or a cognitive one. I submit that, with increasing reliance on technology, our reactive control functions will make us more likely to choose cognitive tasks instead of physical ones.

Some very good and relevant insights about the value or otherwise of cognitive *vs* physical activities are discussed by Alistair Nunn, Research Centre for Optimal Health, Department of Life Sciences, University of Westminster. These are presented in the Addendum below (Nunn A. 2016 Personal communication).

Apart from physical exercise, we may also need to redefine what is 'healthy' with regards to nutrition. We should begin to consider foodstuffs which, although currently considered unhealthy are, in effect, of primary importance within a cognitive environment. For instance, Brickman *et al.* [34] have shown that cocoa flavanols (from chocolate) may be better for neural functioning than exercise. In this study, those who exercised physically did not experience a considerable benefit in cognitive abilities, whereas those who only used the cocoa flavanols did. Apart from chocolate, glucose may also be currying an unfair burden as an unhealthy food. The presence of glucose can enhance the function of neurons, and it is essential during fast neuronal network oscillations. Glucose deprivation adversely affects gamma oscillations and, although energy provided by other energy-rich substrates can also fuel the process of oscillations, glucose is the most effective [35]. Gamma oscillations are involved in information processing and memory formation, therefore it is essential in maintaining a good function of neuronal activity [36]. The authors of this study said:

"This supports the hypothesis that highly energized fast-spiking interneurons are a central element for cortical information processing and may be critical for cognitive decline when energy supply becomes limited ('interneuron energy hypothesis').

Therefore, one simple way of presenting this argument, is that glucose intake plus a boring, sedate lifestyle is bad for health, whereas glucose intake plus physical and/or mental stimulation is good for health. These are examples of how foodstuffs which we are generally advised to avoid, may be very useful in improving cognitive function and make us better able to operate effectively within a technological environment. This environment is now beginning to take form as a new eco-system an amalgam of humans with the internet, termed the 'Global Brain'.

Ageing and The Global Brain: Technologically-Inspired Challenges

The Global Brain (GB) has been defined as a developing distributed intelligence emerging from the interactions of humans and computers/internet [37]. It is an integration of Ambient Intelligence with human agents. As I have argued extensively in this book, humans are increasingly becoming integrated with technology, and particularly, digital communication technology (including the internet). This integration and the consequent crucial role of humans within a global superorganism [38] may have an unprecedented impact on our biology by invoking hitherto dormant evolutionary processes. An important consequence of a human-computer integration is that it may become thermodynamically uneconomical for humans to continue evolving through natural selection – new ways of human evolution will have to emerge. Here, **I am not referring to evolution in the Darwinian sense (the change in the heritable traits of biological populations over successive generations). Instead, I am referring to a more general interpretation of the term 'evolution' which is 'the adaptation in changes to the environment, so that survival continues'.**

It may make more sense from the allocation of resources point of view, to maintain existing integrated humans indefinitely, rather than eliminate them through ageing and generate new ones, who would then need extra resources in order to re-integrate themselves within the GB. The net result will be that we will start experiencing an unprecedented declining of age-related degeneration, with resulting increasing prolongation of our lifespan, during a process whereby the GB evolves to higher levels of complexity at a low thermodynamical cost. This is an indirect consequence of an increasing exposure to meaningful actionable information which has a global impact. Certain basic concepts underlying such a possibility have been discussed in chapter 7.

As a general concept, humans within the GB may be compared to neurons within a brain: it is known that some new neurons are formed during adulthood, at least in certain parts of the brain [39]. This can be analogous to new babies being born to replace any human losses within the GB. However, the majority of cortical neurons are maintained in good operating condition and remain the same entities throughout life, instead of actively being replaced every few weeks (as in the case of, for instance, skin or blood cells). Therefore, we see that well-connected and information-laden neurons live much longer than other somatic cells which have lower information content. These neurons may even outlive their human host [40]. Thus, in this analogy, humans who have low information content (those who are experiencing continual decrease in their functional complexity) will not be maintained for long. By contrast, those who have increased information content will experience the same natural retention mechanisms seen in long-lived neurons, and thus live longer. According to some predictions, humans will increasingly embed themselves within the Global Brain by way of highly sophisticated digital interfaces (first examples are iphones) that can anticipate the subject's wishes, preferences, habits and other parameters. Eventually, there could be suitable technology that can allow direct brain to brain communication [41]. If this becomes the case, it may reasonable to speculate that it will cost more in energy terms to replace a well-integrated human (through creating a new one following conventional processes of somatic ageing and death) rather than maintain the existing one (through improved somatic repair). Returning to the analogy of the brain, research shows that new neurons that are not well-integrated into the brain die prematurely [42, 43]. Therefore, then general principle seems to support the view that, in order to last longer within a network, one needs to be well integrated within that network.

The sequence of events will happen naturally, based on underlying natural laws and principles. Human brains as individual units of the GB, will be subjected to increased pressures in order to survive longer. This is not a teleological argument. The GB does not have any intent or purpose. It is merely an instrument of nature, originating from fundamental events, forming part of the general direction of evolution: from simple to complex [3]. By following a purposeful commitment to embed oneself into the GB and increase meaningful usage of cognitive information of sufficient magnitude, there may be (there **has** to be) up-regulation of somatic repair mechanisms which effect epigenetic changes that diminish age-related degeneration of function, as discussed throughout this book.

Therefore, the incidence and prevalence of age-related degeneration will have to be reduced, in order for the above series of events to take place. If age-related degeneration is not reduced it would be contrary to the general thrust of evolution and also contrary to entropic principles [44].

Stine-Morrow *et al.* [45] have shown that being well- embedded within a complex environment (in the intellectual and social sense): ..."*can impact cognition, perhaps even broadly, without explicit instruction*". This is to say that people who are involved in creative problem-solving situations, such as those encountered in our intensely information-rich, developed society, show significant improvement in divergent thinking and are thus better able to adapt to the environment they are in. An enriched environment (*i.e.* a task environment) provides both resources and challenges which promote up-regulation of our adaptive processes and make us more resilient to stress. This is an essential characteristic of successful ageing (here taken to mean a continued development without any degeneration, and without chronic, time-dependent dysfunction). Therefore, it could be argued that exposure to information such as *via* the internet, may positively influence health and longevity.

There is empirical evidence that these concepts are valid within human groups in the community. D'Orsi *et al.* [46] have shown that internet use (as well as physical activity) were significantly associated with lower risk of 10-year mortality, controlling for sex, age, education, cognitive function, functional impairment and diabetes. This is one of the few studies to show a positive correlation between internet use and health in later life. The researchers used quantitative regression analysis to examine the relationship between internet use and life expectancy. The found that there is a significant such correlation which does not solely depend on the economic condition of each country. Although there seems to be a slowing down of the increase of life expectancy after a certain number of internet users has been reached, a closer analysis using the Pearson correlation coefficient shows that there is a definite increase of average life expectancy with each unit of internet use increase, from 0.537 years per unit of internet use in 1999 to 0.766 years per unit of internet use in 2010 [47]. One may well interpret the results to be due to an improvement of biological function, following up-regulation of stress response after a challenging exposure to hormetic cognitive information *via* the internet.

One very valid question regarding the Global Brain notion and human longevity, is how to show that there are feedback mechanisms which either maintain or allow participant humans to perish. It is important to emphasise that the Global Brain is nothing more than an abstract construct, a combination of humans and the internet. It has no reasoning powers and cannot spontaneously decide what action to take. It is bound by basic laws and principles grounded in elementary physics, cybernetics and computer science. Therefore the feedback mechanisms are buried deeply in these principles and it is not a matter of simply anthropomorphising the Global Brain which has to 'decide' one way or another. One possible feedback mechanism is selective reinforcement which has been discussed elsewhere in this book. Agents (humans) within the Global Brain are selectively identified and allowed to continue functioning if their contribution is beneficial to the overall system. Developing valuable links with other individuals, results in either stimulation or boredom. The stimulating links are preferentially selected, and this means that the individual is now more likely to experience a positive stimulation and less likely to experience boredom. This improves both psychological well-being and physiological parameters (improved cognitive mechanisms and processes, and increases the information-content of the individual). The concept of the Global Brain is absolutely necessary because it considers humans as being part of a self-organising network, and not in isolation. There are hidden benefits based on emergence: the appearance of effects which cannot be reduced back to individual components of a system. Therefore suitable integration within a self-organising, complex amalgam of humans and computers is a necessary step towards improved longevity.

Ambient Intelligence

Within the concept of the Global Brain, it may be useful to also examine the notion of Ambient intelligence (AmI). This will help clarify certain aspects of digitally-based environmental challenges. AmI refers to a situation where there is widespread use of computing devices, where the physical and technological environment interacts with humans [48]. The information-processing capacity is embedded in our environment in ways we may not even be aware of. This type of informational landscape is becoming progressively more widespread and it is important to consider its effects on human biology, health and behaviour. In addition, it is bound to have effects on basic biological processes, stress response pathways, repair mechanisms and resource allocation priorities. The intelligence which is included in the environment (AmI) is usually distributed, meaning that it is spread-out and divided between different computer devices, in the home, office, school, shops, hospitals *etc*. This intelligence empowers us to access useful information which we may use in order to improve our health, productivity, and general well-being [49].

Ambient Intelligence (AmI) is an adaptive process which is sensitive to new information and responsive to the needs of its users. One emerging expression of AmI is Intelligent Mixed Reality (IMR), which is an ideal blend of virtual reality and physical reality [50]. The system is able to observe and learn from the user, and display scalable intelligence. It takes into consideration social and emotional factors. This type of intelligence does not only influence the human user but it also acts on its environment, again indirectly influencing the user. The user, reciprocally helps the intelligent system adapt and improve: a process of 'co-evolution' which further facilitates human adaptation and advancement. Our environment is no longer a passive ambient but it is actively self-organising, monitors our actions and reflects our behaviour. This increasingly pervasive human-computer integration is having a profound impact on our biology, as it involves an accelerated pace of information-sharing between our neural tissues and the external technological environment. The information we exchange with others such as videos, texts, emails, graphics and sound, is a new form of environmental stimulation (challenge), a potential hormetic experience, which affects our stress response pathways. In the case of AmI we see that this can be useful in two ways:

1. To empower us to access health-related information and use it to improve our health
2. To act as a hormetic challenge which up-regulates our repair and defence mechanisms

The concept of 'smart cities' is also relevant here [51]; see Chapter 6, (Fig. **1**). In a smart city there is integration of conventional operations and networks with digital and virtual tasks, which improves the function of its inhabitants (as well as, in a more abstract sense, improves the function of the city itself). A smart city is a 'self-aware' complex adaptive system, with embedded sensors (the Internet of Things) which continually monitor functions, performance, needs and other characteristics of the entire city. It is clear that humans which are embedded within such a system of reciprocally-influencing feed-back loops, cannot be allowed to perish easily (through ageing) as this would cause a considerable damage to the evolving network [52]. Such networks evolve according to natural principles and have no reasoning or deciding power. If an element of a network is not contributing to the system or if it is damaging to the system, then it will be selectively eliminated. This is based on simulated annealing (https://en.wikipedia.org/wiki/Simulated_annealing): A network is always seeking optimal solutions and this means that negative elements within that network will not be retained. The exact mechanisms that are involved in the retention or in the elimination of an agent within a network need further study and clarification. Nevertheless, the general concept appears to be valid in all systems and at all levels studied so far.

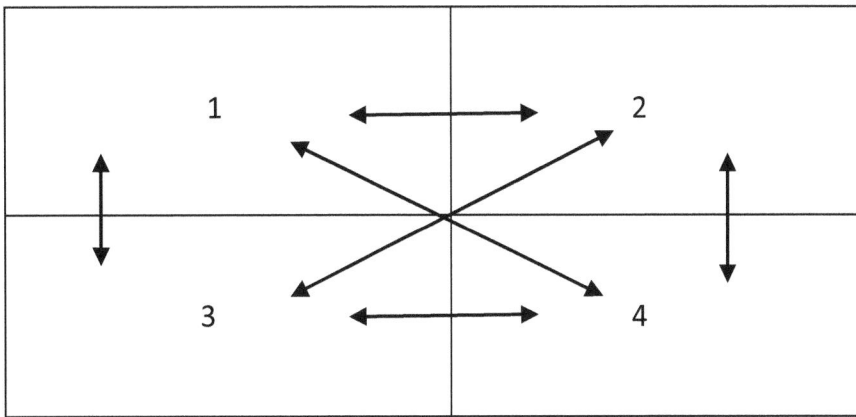

Fig. (1). Interrelationship between the four domains of how internet (technology) use affects life-span extension. Each domain influences and is being influenced by all others.

It can thus be argued with some confidence that a technological environment consisting mostly of information-rich internet-based activities, may help increase human lifespan. There is a number of ways this may occur:

1. *Practical/applied.* This is based on knowledge creation and dissemination of health information. People use information mined from the internet in order to improve their health [53]. This includes any other type of health-related activities, such as governments using health data in order to improve health policies (for instance [54]:), doctors using the internet in order to improve patient care [55, 56], and researchers using informatics in order to devise suitable treatments [57].

2. *Social/Cultural*. Technological and internet-based developments reduce redundant and repetitive work by creating labour-saving tools [58]. This then

allows increasingly more people free time to connect online and better chances of improving their creativity and cognitive abilities. It may also become possible for humans to be **physically** protected by the technological society: for instance, intelligent sensors which are able to diagnose emergency, acute or subacute medical conditions, such as atrial fibrillation, myocardial infarction *etc.*, and then trigger a series of actions that can save the person's life (for examples see: [59]. Thus, the more deeply-embedded one is within the technological society (the GB), and the more technological devices or facilities one uses, the higher the chances of that person being diagnosed, treated and thus survive longer.

3. *Evolutionary*. The use of the internet by many people creates intellectual capital (large groups of well-connected people) which can act as a catalyst for further knowledge creation [60]. Society can then adapt to the increasing knowledge, which provides further scope of development, improvement of the human condition and more creativity [61]. In addition, technology may facilitate the process of a transition from evolution by natural selection, to intentional evolution. For instance, Mahmoudi *et al.* have shown that exposure to electromagnetic radiation from Wi-Fi routers may contribute to male infertility [62]. Furthermore, use of laptops is also associated with reduced male fertility [63]. Although this may appear worrying at first, it may be an instance where technology is causing a shift from high reproduction and shorter lifespans (*r*- selection model), to a society where reproduction is reduced, with necessarily longer lifespans (the *K*- selection model) [64].

4. *Biological.* The sheer impact of information and knowledge humans are exposed to, creates new biological situations and opportunities, which may influence repair processes and resource allocation mechanisms [65]. This may ultimately up-regulate our biological function directly, and cause biological transitions as discussed in this chapter. There is another possible way by which we may derive benefit from living in a technological environment. The 'radiation smog' created by devices such as mobile telephones, Wi-Fi *etc.*, may initiate hormetic effects, which then modify several health parameters. For instance, it was shown that exposure to low dose radiation may help rescue living neurons from degeneration, irrespective of the mechanism of degeneration involved [66]. Much more research is needed in order to describe

these notions in detail, but at least we have made a start.

All of the above mechanisms share common frontiers and are mutually cooperative and reciprocally influencing (Fig. **1**).

Applying a Business Model in Ageing Management

In order to view the matter of the environmental impact on homeostasis in ageing from a different angle, it can be useful to discuss health and ageing using concepts borrowed from business models. This is another link in the tangled web of concepts involved in ageing; learning basic universal principles from diverse areas, including business (Fig. **1**, chapter 6). In business, there are four zones which can be envisaged as representing four different states of **intensit**y of energy and **quality** of energy. The balance between these two characteristics determines one of four situations [67].

A. The Comfort Zone

In business, this is a stage where the environment is stable, calm and secure, with low levels of alertness, low numbers of new initiatives and progressive inertia and inaction. When there is a significant perturbation of the system (a challenge) then there is an inability to react effectively and a difficulty in adapting to the new stimulus. This results into lack of progress and poor survival of the company. In biology, this situation can be compared to a stage where the biological mechanisms are functioning efficiently at low levels, without any effort to change the *status quo,* producing adequate amounts of compounds which keep the body just right, within homoeostatic constraints. But when the system is challenged, there is an inability to restore homoestasis and the system cannot adapt, increasing its chances of degeneration and early death.

In clinical and social terms, this is the equivalent of an average existence, dealing with life problems as they appear, being contented with what life has to offer. This is an evolutionary *status quo* which allows for disease and ageing to develop according to existing evolutionary dynamics (see chapter six, Fig. **5**, the 'A' phenotype).

B. The Resignation Zone

Here the energy levels are comparable to (or worse than) the previous zone. In business, there are high amounts of negative emotions such as frustration, detachment, poor confidence and low alertness. The system functions at a mediocre level and is unable to cope with any significant problems or perturbations. In biology this can be the equivalent of **dysfunction** with biological mechanisms functioning at a low, suboptimal, levels and being easily subjected to malfunction. Any unexpected threats, challenges or stresses reinforce the increasingly bad state of biological functionality. There is routine, predictability and monotony in biological signals, instead of purposeful, dynamic activity. Clinically, the Resignation zone reflects subclinical disease, and poor emotions such and anxiety and depression. Individuals passively resign to their fate and are unable or unwilling to react. Routine, apathy and boredom are high determinants for disease.

C. The Corrosion Zone

Still associated with dysfunction, the Corrosion Zone is, in a business model, a zone of high emotions, misplaced energy and intense activity which is over and above what the organisation can tolerate in the long term. Despite high levels of activity, people remain unfocused. The competition levels are high resulting in lack of clear progress. In the biological realm this is the equivalent of an exaggerated stress response, with cells and signalling pathways overacting and overworking heading towards burnout. It becomes less likely that the system will be able to return within the 'safe heaven' of the homeodynamic space. This is reflected in the clinical realm with debilitation and obvious signs of stress, disease and eventual lack of function. In this respect, age-related degenerative disease can thrive and the chances of accelerated ageing are increased.

D. The Productive Zone

The Productive zone is characteristic of high levels of **focused** activity and alertness, drive and enthusiasm. There is cooperation instead of competition. Biologically, this is a level where any external challenges are met successfully, with the system adapting and improving compared to a stage before the challenge.

The system is able to cope with a relatively large amount of challenges (but which do not exhaust its resources), with the resources growing in tandem with the degree of challenging demands. In other words, the system evolves successfully. The energy levels are increased and the information flow between the different components escalates towards an ideal level. Clinically this is a zone of increased alertness, a search for novelty and new opportunities, where hormesis and enrichment are rife. This zone is ideal for health and lifespan extension. The aim of this book is to encourage this situation and promote strategies which can be useful in attaining this stage.

So people make their own choices (or allowed to gravitate towards bad choices) and the result reflects the above general concept. If one is satisfied with living their life in mediocrity or in attrition with others then they will be unlikely to persist (survive) long. If they purposefully step out of their comfort zone in all respects and choose a situation of productive (in business) or increased positive challenges (in biology/clinical) then these individuals are more likely to live longer and function for longer (as in chapter 6, Fig. **5**, the 'B and C' phenotypes).

Continuing this short study of the business model analogy in ageing, it is possible to discuss how to overcome 'non-action', and move towards purposeful activity – which, as mentioned above, is more likely to improve usefulness, and thus longevity:

- Map a realistic worldview (develop a list of ideas or objects you want to achieve)
- Map a realistic time-structure for achieving the above
- Plan for any possible problems, bottlenecks and constraints
- Accept and manage trade-offs (for example accept that you cannot remain healthy without making any effort on your behalf, and also use technology to overcome any physical trade-offs against cognition)
- Accept that you need to be resilient in the face of a challenge or adversity
- Accept that you need to explore new concepts and choices

This can be achieved through a *'slaying the dragon-winning the princess'* parable:

1. A 'slaying the dragon strategy' is based on driving individuals out of their

comfort zone by focusing on a crisis or a danger. In this case, one can move out of their comfort zone if they focus on the terrible consequences of age-related degeneration and death.

2. A 'winning the princess' strategy on the other hand, is to concentrate on a worthwhile prize to match the effort, creating a vision in order to improve motivation. In the case of ageing, this would be to concentrate on the benefits of a life without age-related degeneration.

3. A combination of the two.

Without any focus on danger (1) or focus on benefit (2) the individual will gradually decline and easily experience all the dysfunction elements of ageing.

CONCLUSION

The transition from evolution by natural selection to a post-Darwinian domain (where we become able to influence our own evolution) is one of the most important transitions in biology, and unique in human history. We are beginning to witness a true Kuhnian paradigm shift, a change in fundamental sets of beliefs and assumptions about human evolution (as defined in the glossary in chapter 1). The anomaly of death by ageing cannot be explained by modern emerging scientific doctrines (self-organised systems, hierarchically emergent complex neural, digital, technological and societal structures). The new paradigm may explain why, in order for further evolution of human intellectual sophistication, it is necessary for (a significant number of) humans to remain alive for longer, without experiencing age-related degeneration. These approaches have practical significance now, and are available to anyone, without the need to wait for any future artificial treatments. There is no point in considering these activities in isolation. It is the totality of these strategies which is based on a reciprocally-influencing principle that matters. The activities have a direct anatomical and biological impact upon hormonal, immune, neuronal and other processes that are valuable in repairing and preventing age-related damage. In addition, the practice of challenge exposure as a whole is aided by natural evolutionary processes that enhance fitness and survival within a technological niche. It is predicted that, in order for the process to be successful, it is not a matter of how well an individual is enriched but also a matter of how many enriched individuals interact with each other. In other words the elimination of age-related degeneration is constrained by the interconnectedness of a critical number of appropriately challenged individuals. Therefore it is the responsibility of all of us to work together in order to achieve a life without the chronic dysfunction of ageing.

A request: the reader is requested to revisit the master figure in the Preface and review once again the different components of the ideas discussed here. This will create a more holistic understanding of the concept, and it may help place any apparently unrelated ideas into one collective frame.

Addendum (Alistair Nunn, 2016 Personal communication)

"Could cognitive effort be superior to physical effort in relation to controlling inflammation"? That is indeed the billion dollar conundrum. I think the honest answer here is that we simply do not have enough data to empirically answer it, and part of the reason for this is that we still don't fully understanding exactly what "thinking" is. The implication is that the right kind of thinking could offset a lack of physical activity as a way to suppress inflammation and thus the ageing process. I do make a comment in relation to this about Stephen Hawking (chapter 9). However, and I say this fairly seriously, we need to tread very carefully here because the data is extremely strong to support the idea that increasing physical activity is highly beneficial, in particular, as it reduces inflammation, but the thinking concept is not. This is not to say it is not important, but in a world that struggling because it is getting so sedentary, the last thing we want to do is to provide yet another excuse for people not to exercise! In a word, we have a duty of care, so need to be careful how we portray these ideas.

If, for the time being, we follow the logic we have developed in this paper (chapter 9), then what appears to be determining the ageing process is an energy/informational balance that expresses itself as inflammation. If this process is centrally controlled by the brain, for instance in the hypothalamus, then one could make the argument that constantly engaging/stimulating the brain would signal that suppressing inflammation is important for survival as adaptability of the individual is paramount. That part, I think, makes perfect sense. However, where it becomes murkier is that as engaging the brain is key in survival, and that usually means maintaining a good energy supply ("cognitive buffering"); being smart helps you find/grow your own food. This is the thermodynamic point we make about energy flow determining replication and thus, modulating survival of the individual. Hence, in a plentiful environment, this process is simply suppressed and is apparently beyond our control due to basic quantum/thermodynamic considerations. The clearest example of this is that the size of fat cells is limited, and when they get too big, they send out all sorts of inflammatory signals; data now show that even if you are fat and fit, you still don't live as long compared to someone who is of a healthy weight and fit. Now, the interesting point here is that inflammation normally suppresses appetite, suggesting a negative feedback mechanism – indeed, it has been suggested that diabetes is actually the bodies' final attempt to stop the fattening process – as this clearly starts to have a negative impact on survival (the organism cannot move quickly enough to escape predators, or possibly, fight for mates); this may even extend to food-induced thermogenesis, which apart from being a way to reduce ROS by uncoupling, may also serve a secondary purpose to burn of excess calories. So here is the point: the only way that cognitive activation might suppress inflammation in a positive-calorie sedentary environment (where the propensity is always to get fatter and the amount of muscle to reduce), where thermodynamics will drive inflammation (cells have a finite size and thus have to replicate), is either to reduce intake, or accelerate energy use, or, and this perhaps what you are after, to suppress inflammation – which we know can be done centrally (*e.g.*, *via* activation of the parasympathetic nervous system). But the rub of the latter point is that you would simply continue to get fatter. To flip this, it is clear that there are many people, despite the obesogenic environment, who do manage to stay thin and fit even when they could easily become fat because they are well off; they generally live a lot longer than people who "let themselves go". One of the reasons for this, which we put forward in one of our earlier papers, is that there exists a "tipping point", where inflammation suppresses the will to exercise and the system enters into a kind of negative cycle, because the inflammatory suppression of appetite does not work. In evolutionary terms, this makes perfect sense, as inflammatory suppression of movement is key to allow an injury to heal or an infection to be overcome, but appetite must return if the animal is to survive, as not eating is a dead end evolutionary strategy! In effect, in a modern world, humans live in a perpetual state of mild inflammatory suppression of exercise, but not appetite. Which of course is exactly what you would predict form the thermodynamic principles we outline in this new paper. Thus the conclusion is that there is a sub-population of long-lived humans in an obesogenic world who seem to thrive by actively engaging in exercise and managing their intake, and interestingly, the evidence seems to point to this being associated with a higher IQ, or at least, a higher social class. Cause, effect, or self-determinism – or intriguingly, something else? Nature *versus* nurture?

So where does this leave us? Does having an improved (genetic?) capacity to think lead to one living longer, or is it something one can learn? Is there a subpopulation of people who are extremely fat, but very bright, who live a long time? The point here is that if you expand that mitochondrial/nuclear DNA matching theory, it could be argued that the capacity for thought/intelligence is directly linked to longevity and mitochondrial efficiency, which is the flip side, to some degree, of the propensity to inflammatory loss of structure. The fat person who lives a long time may do so because they simply have more efficient mitochondria that enables them to tolerate high levels of fat laden cells, and they don't require high levels of muscle to provide the anti-inflammatory signal. The interesting question is therefore whether or not forcing the brain of an average fat person to think can increase the efficiency of mitochondria throughout the body as an adaptive response, possibly in much the same way as the reverse happens when the muscles are made to work putting metabolic demands on the rest of the body, including the brain. We do touch upon this in relation to meditative practices, as it may well be that they force the brain into a more coherent state, which by definition, would have to enhance mitochondrial efficiency if practiced often enough. But is this process as efficient as making someone exercise? My first thoughts would be no, for the very simple reason that exercise enforces a whole body adaptation that involves not only the muscles, but the liver and other organs required to deal with the metabolic load, and particularly, the brain, as it has to coordinate everything, and results in a massive output of calories that would offset intake; in effect, exercise is a kind of enforced hormesis as the body has no choice – and has been happening throughout evolution. On the other hand, evolution has rarely, if ever, given the luxury of simply sitting still and getting fat while thinking – and that is a voluntary choice; however, throughout evolution we were often forced to think in order to survive, but even here, it is tightly controlled to ensure we don't over tire the brain.

My sense is that thinking is hormetic, and can certainly, I believe, have a beneficial effect – especially if combined with exercise. Both yoga and Tai chi emphasise the important of "being in the moment" when training, and focussing completely on the movements while relaxing; this concept has been tested throughout history because it works. But for me, one of the most important aspects is simply energy flow: even when we think very hard, our overall neural energy use does not go up that much (10-20%, although it is pretty clear that the science has not got up with this yet, so these figures are not that accurate). Hence, I think the message must be that it can certainly help, but can it ever be a replacement for exercise? I don't think so, and even if it could, it would require immense discipline to do. It is much easier to measure how far you have run, rather than how hard you have thought – unless we could come up with a very accurate way of measuring energy use in the brain (which is not possible at the moment). The other point is that movement does require cognition, in particular, new movement. One example of the benefits of this relate to balance, which is a very good indicator of overall brain-muscle integration. One of a new series of measures to indicate future mortality is simply to measure how well someone stands up from a cross-legged seated position. This is why so many practices put emphasis on "mind-body" unification.

Again, we must look to basic thermodynamics and evolution, and to a degree, the wisdom already inherent in society generated through millennia. What does seem to be the case is that putting one's awareness into a particular part of the body, say, a muscle does induce changes (for example, if, in a trial, you immobilise an arm of one group of volunteers and leave it for a few weeks, the muscles atrophy, but if you tell another group with the same immobilisation to put their mind into the arm and imagine it moving, the muscle loss is much less). This idea of "intent" is very real, suggesting that the brain does have a powerful effect, well, we know it does from the placebo effect. If I had a research budget, one of the first things to test would be to show that thinking can alter say, muscle or immune mitochondrial function, so providing a real solid basis for the theory – and then develop a feedback mechanism for the practitioner to enhance it. However, I suspect that we can uncouple the movement from the brain, and the brain from the movement, but the best effects will only happen when combined. In case you are wondering, I study both Tai chi and yoga, and used to do aerobatics in gliders; visualisation of the routines before you took off was vitally important. Visualisation is key in many sports, and I think this shows that integrating thinking and movement is something selected for by evolution. But, and here is the thing, laziness is also an incredibly important trait as it saves energy, and has been heavily selected for, but it is offset by the propensity for humans, and other animals, to show curiosity and self-engagement to solve problems – we seem to enjoy this, which suggests it is an evolved trait. As soon as it seems like work we avoid it! This could partly explain the intelligence related longevity idea; if people are engaged in something they enjoy, and don't feel like they are being managed, they tend to be self-incentivised – as long as they earn enough to be comfortable. In effect, being creative is the best state to be in (rather than being forced to repeatedly do the same routine thing, where only external incentives seem to work). Thus, the problem we have with today's society is how to self-incentivise people by engaging their creativity, so they willingly stimulate their brain; sadly, many corporations work *via* having "human resources", effectively seeing each employee as a profit-delivery unit requiring management, external incentives, which stifling creativity. It is not surprising that many folk view any form of exercise, whether it be physical or mental, as another form of enforced routine to be avoided.

So I think that hormetic thinking could tip the brain towards an anti-inflammatory state, which would unleash its ability to self-incentivise, and enable enjoyment of exercise, which would provide the metabolic hormesis required to complete the process. In evolution terms, one could then see how this would engender a positive feedback survival loop, as not only would the animal become fitter & smarter, it would drive it to seek more food and eventually reproduce. Equally, if inflammatory tone rises, which it does as we age, one can also see how a negative survival loop would start to be initiated that would lead to its eventual death. The key factor here is "purpose"; the need to survive to breed provides a real purpose, but in humans, it seems that recognition and mastery of something is its own reward – but again, in evolution terms, this makes perfect sense. I guess the point I am making here is that hormesis provides "purpose" as it is a form of stress requiring adaptation, as defined by basic thermodynamic rules. What I guess this boils down to is that in order to stimulate the brain, it requires purpose, and this is dictated by the environment and the inflammatory state; movement is a very powerful anti-inflammatory agent. If you don't need to move, then there is not much purpose to life other than to breed, as it indicates a non-challenging environment, which, throughout evolution, was really, if ever, the case. So while thinking can be hormetic, I suspect that our genes have evolved around the need to move, as not moving was generally fatal for a number of reasons – but this had to be coupled to the ability to control it. Which of course begs the question, at what level of complexity did "thinking" evolve? Or should we not call it thinking at all, but the process of integrating "memory" with the "now", to "predict" the future to survive – something that all orders of life seem to be able to do – and which is tightly coupled to energy flux. It does certainly appear to be the case that many animals do "better" in an interesting and engaging environment; we need to do the experiments to prove it. Finally, there is evidence that bioactive polyphenols can "fool" the system into thinking it is going into a negative energy state, so slowing the ageing process; but even here, there is probably a strong evolutionary reason due to the concept of "xenohormesis" – so it is not fooling it all, as an increase in bioactive compounds simply heralds the beginnings of a difficult time (*e.g.*, autumn). Hence, conditions that stimulate the brain are hormetic because they herald the need to think, or at least process information, to survive. So is it possible that "smart" people tend to survive longer because they find it easier to "think", and this has the ability to not only suppress inflammation directly, but modulate appetite more effectively and increase exercise levels, so maintaining an ideal weight and thus fitness.

In summary, I think that thinking could indeed offset some of the downsides of a sedentary existence, but not all of it, because thinking and movement have been intimately connected through necessity throughout evolution. This may well be modulated by both levels of fat and muscle, inflammatory status (injury, both physical and mental, and of course, infection) and quite possibly, innate cognitive ability;

some people will finder it easier than others. The danger is that some kinds of thinking are clearly more hormetic than others, and because it is almost impossible to measure, it would be a lot more difficult to monitor than exercise. The brain clearly monitors energy stores, and in times of hardship, will obviously be prioritised (*e.g.*, insulin resistance). I guess the ultimate point here is that it is a question of energy flux; thinking does not metabolically load the body as much as exercise, although it can prepare the body for harder times (cognitive hormesis) – having to exercise in a calorie restricted cold environment is a very, very powerful and immediate stimulus as it rapidly induces negative calorie balance – probably far more so than thinking about it. So until we know for sure, I think we should hedge our bets a bit.

REFERENCES

[1]   Kyriazis M. Practical applications of chaos theory to the modulation of human ageing: nature prefers chaos to regularity. Biogerontology 2003; 4(2): 75-90.
        [http://dx.doi.org/10.1023/A:1023306419861] [PMID: 12766532]

[2]   Stewart JE. The direction of evolution: the rise of cooperative organization. Biosystems 2014; 123: 27-36.
        [http://dx.doi.org/10.1016/j.biosystems.2014.05.006] [PMID: 24887200]

[3]   Sleimen-Malkoun R, Temprado JJ, Hong SL. Aging induced loss of complexity and dedifferentiation: consequences for coordination dynamics within and between brain, muscular and behavioral levels. Front Aging Neurosci 2014; 6: 140.
        [http://dx.doi.org/10.3389/fnagi.2014.00140] [PMID: 25018731]

[4]   Kyriazis M. Applications of chaos theory to the molecular biology of aging. Exp Gerontol 1991; 26(6): 569-72.
        [http://dx.doi.org/10.1016/0531-5565(91)90074-V] [PMID: 1800131]

[5]   Lee SL, Thomas P, Fenech M. Genome instability biomarkers and blood micronutrient risk profiles associated with mild cognitive impairment and Alzheimer's disease. Mutat Res 2015; 776: 54-83.
        [http://dx.doi.org/10.1016/j.mrfmmm.2014.12.012] [PMID: 26364206]

[6]   Xu B, Jin G, Ma Y. Evolution kinetics and phase transitions of complex adaptive systems. Physical Rev 2005; E71: 026107: 1-5.
        [http://dx.doi.org/10.1103/PhysRevE.71.026107]

[7]   Michod RE, Nedelcu AM. On the reorganization of fitness during evolutionary transitions in individuality. Integr Comp Biol 2003; 43(1): 64-73.
        [http://dx.doi.org/10.1093/icb/43.1.64] [PMID: 21680410]

[8]   Pollack GH, Wei-Chun C, Eds. Phase Transitions in Cell Biology. Netherlands: Springer 2008.
        [http://dx.doi.org/10.1007/978-1-4020-8651-9]

[9]   Scott A. Characterizing phase transitions in a model of natural evolutionary dynamics. Bull Am Phys
        Soc 2013; 8: 1.

[10]   Anderson JC, Laughlin SB. Photoreceptor performance and the co-ordination of achromatic and
        chromatic inputs in the fly visual system. Vision Res 2000; 40(1): 13-31.
        [http://dx.doi.org/10.1016/S0042-6989(99)00171-6] [PMID: 10768038]

[11]   Kann O, Papageorgiou IE, Draguhn A. Highly energized inhibitory interneurons are a central element
        for information processing in cortical networks. J Cereb Blood Flow Metab 2014; 34(8): 1270-82.
        [http://dx.doi.org/10.1038/jcbfm.2014.104] [PMID: 24896567]

[12]   Rajan K, Bialek W. Maximally informative "stimulus energies" in the analysis of neural responses to
        natural signals. PLoS One 2013; 8(11): e71959.
        [http://dx.doi.org/10.1371/journal.pone.0071959] [PMID: 24250780]

[13]   Engl E, Attwell D. Non-signalling energy use in the brain. J Physiol 2015; 593(16): 3417-29.
        [http://dx.doi.org/10.1113/jphysiol.2014.282517] [PMID: 25639777]

[14]   Stewart J. The Self-Organizing Society: A Grower's Guide 2015. Available at SSRN:
        http://ssrn.com/abstract=2657948 (accessed 26 November 2015)
        [http://dx.doi.org/10.2139/ssrn.2657948]

[15]   Barone I, Novelli E, Strettoi E. Long-term preservation of cone photoreceptors and visual acuity in
        rd10 mutant mice exposed to continuous environmental enrichment. Mol Vis 2014; 20: 1545-56.
        [PMID: 25489227]

[16]   Dorfman D, Aranda ML, González Fleitas MF, *et al.* Environmental enrichment protects the retina
        from early diabetic damage in adult rats. PLoS One 2014; 9(7): e101829.
        [http://dx.doi.org/10.1371/journal.pone.0101829] [PMID: 25004165]

[17]   Gurfein BT, Davidenko O, Premenko-Lanier M, *et al.* Environmental enrichment alters splenic
        immune cell composition and enhances secondary influenza vaccine responses in mice. Mol Med
        2014; 20: 179-90.
        [http://dx.doi.org/10.2119/molmed.2013.00158] [PMID: 24687160]

[18]   Vitalo AG, Gorantla S, Fricchione JG, *et al.* Environmental enrichment with nesting material
        accelerates wound healing in isolation-reared rats. Behav Brain Res 2012; 226(2): 606-12.
        [http://dx.doi.org/10.1016/j.bbr.2011.09.038] [PMID: 22008380]

[19]   Zhao Y, Chen K, Shen X. Environmental Enrichment Attenuated Sevoflurane-Induced Neurotoxicity
        through the PPAR-γ Signaling Pathway. Biomed Res Int 2015; 2015: 107149.

[20]   Matur E, Akyazi İ, Eraslan E, *et al.* The effects of environmental enrichment and transport stress on
        the weights of lymphoid organs, cell-mediated immune response, heterophil functions and antibody
        production in laying hens. Anim Sci J 2015. [Epub ahead of print].
        [http://dx.doi.org/10.1111/asj.12411] [PMID: 26419323]

[21]   Cai Y, Abrahamson K. How exercise influences cognitive performance when mild cognitive
        impairment exists: a literature review. J Psychosoc Nurs Ment Health Serv 2015; 18: 1-11. [Epub
        ahead of print)].

[PMID: 26565414]

[22] Clark BC, Mahato NK, Nakazawa M, Law TD, Thomas JS. The power of the mind: the cortex as a critical determinant of muscle strength/weakness. J Neurophysiol 2014; 112(12): 3219-26.
[http://dx.doi.org/10.1152/jn.00386.2014] [PMID: 25274345]

[23] Saucedo Marquez CM, Vanaudenaerde B, Troosters T, Wenderoth N. High intensity interval training evokes larger serum BDNF levels compared to intense continuous exercise. J Appl Physiol 2015; 18: 1-11. [Epub ahead of print)].
[http://dx.doi.org/jap.00126.2015]

[24] Navarro F, Bacurau AV, Pereira GB, *et al.* Moderate exercise increases the metabolism and immune function of lymphocytes in rats. Eur J Appl Physiol 2013; 113(5): 1343-52.
[http://dx.doi.org/10.1007/s00421-012-2554-y] [PMID: 23212119]

[25] Gurfein BT, Davidenko O, Premenko-Lanier M, *et al.* Environmental enrichment alters splenic immune cell composition and enhances secondary influenza vaccine responses in mice. Mol Med 2014; 20: 179-90.
[http://dx.doi.org/10.2119/molmed.2013.00158] [PMID: 24687160]

[26] Zebunke M, Puppe B, Langbein J. Effects of cognitive enrichment on behavioural and physiological reactions of pigs. Physiol Behav 2013; 118: 70-9.
[http://dx.doi.org/10.1016/j.physbeh.2013.05.005] [PMID: 23680428]

[27] Wolinsky FD, Unverzagt FW, Smith DM, Jones R, Wright E, Tennstedt SL. The effects of the ACTIVE cognitive training trial on clinically relevant declines in health-related quality of life. J Gerontol B Psychol Sci Soc Sci 2006; 61(5): S281-7.
[http://dx.doi.org/10.1093/geronb/61.5.S281] [PMID: 16960242]

[28] Rebok GW, Ball K, Guey LT, *et al.* ACTIVE Study Group. Ten-year effects of the advanced cognitive training for independent and vital elderly cognitive training trial on cognition and everyday functioning in older adults. J Am Geriatr Soc 2014; 62(1): 16-24.
[http://dx.doi.org/10.1111/jgs.12607] [PMID: 24417410]

[29] Wolinsky FD, Mahncke H, Vander Weg MW, *et al.* Speed of processing training protects self-rated health in older adults: enduring effects observed in the multi-site ACTIVE randomized controlled trial. Int Psychogeriatr 2010; 22(3): 470-8.
[http://dx.doi.org/10.1017/S1041610209991281] [PMID: 20003628]

[30] Hochachka PW. Solving the common problem: matching ATP synthesis to ATP demand during exercise. Adv Vet Sci Comp Med 1994; 38A: 41-56.
[PMID: 7801835]

[31] Troubat N, Fargeas-Gluck MA, Tulppo M, Dugué B. The stress of chess players as a model to study the effects of psychological stimuli on physiological responses: an example of substrate oxidation and heart rate variability in man. Eur J Appl Physiol 2009; 105(3): 343-9.
[http://dx.doi.org/10.1007/s00421-008-0908-2] [PMID: 18987876]

[32] Leedy C, DuBeck L. Physiological changes during tournament chess. Chess Life and Review 1971; p. 708.

[33] Kurzban R, Duckworth A, Kable JW, Myers J. An opportunity cost model of subjective effort and task

performance. Behav Brain Sci 2013; 36(6): 661-79.
[http://dx.doi.org/10.1017/S0140525X12003196] [PMID: 24304775]

[34]   Brickman AM, Khan UA, Provenzano FA, *et al.* Enhancing dentate gyrus function with dietary flavanols improves cognition in older adults. Nat Neurosci 2014; 17(12): 1798-803.
[http://dx.doi.org/10.1038/nn.3850] [PMID: 25344629]

[35]   Galow LV, Schneider J, Lewen A, Ta TT, Papageorgiou IE, Kann O. Energy substrates that fuel fast neuronal network oscillations. Front Neurosci 2014; 8: 398.
[http://dx.doi.org/10.3389/fnins.2014.00398] [PMID: 25538552]

[36]   Kann O, Papageorgiou IE, Draguhn A. Highly energized inhibitory interneurons are a central element for information processing in cortical networks. J Cereb Blood Flow Metab 2014; 34(8): 1270-82.
[http://dx.doi.org/10.1038/jcbfm.2014.104] [PMID: 24896567]

[37]   Heylighen F. Conceptions of a Global Brain: an historical review. In: Grinin LE, Carneiro RL, Korotayev AV, Spier F, Eds. Evolution: Cosmic, Biological, and Social. Kirov, Russia: Uchitel Publishing 2011; pp. 274-89.

[38]   Heylighen F. The Global Superorganism: an evolutionary-cybernetic model of the emerging network society. Social Evolution & History 2007; 6(1): 58-119.

[39]   Danzer SC. Adult Neurogenesis: Opening the Gates of Troy From the Inside. Epilepsy Curr 2015; 15(5): 263-4.
[http://dx.doi.org/10.5698/1535-7511-15.5.263] [PMID: 26448730]

[40]   Magrassi L, Leto K, Rossi F. Lifespan of neurons is uncoupled from organismal lifespan. Proc Natl Acad Sci USA 2013; 110(11): 4374-9.
[http://dx.doi.org/10.1073/pnas.1217505110] [PMID: 23440189]

[41]   O'Doherty JE, Lebedev MA, Ifft PJ, *et al.* Active tactile exploration using a brain-machine-brain interface. Nature 2011; 479(7372): 228-31.
[http://dx.doi.org/10.1038/nature10489] [PMID: 21976021]

[42]   Alvarez-Buylla A, Kohwi M, Nguyen TM, Merkle FT. The heterogeneity of adult neural stem cells and the emerging complexity of their niche. Cold Spring Harb Symp Quant Biol 2008; 73: 357-65.
[http://dx.doi.org/10.1101/sqb.2008.73.019] [PMID: 19022766]

[43]   Lennington JB, Yang Z, Conover JC. Neural stem cells and the regulation of adult neurogenesis. Reprod Biol Endocrinol 2003; 1: 99.
[http://dx.doi.org/10.1186/1477-7827-1-99] [PMID: 14614786]

[44]   Sharov AA. Evolutionary constraints or opportunities? Biosystems 2014; 123: 9-18.
[http://dx.doi.org/10.1016/j.biosystems.2014.06.004] [PMID: 25047708]

[45]   Stine-Morrow EA, Payne BR, Roberts BW, *et al.* Training *versus* engagement as paths to cognitive enrichment with aging. Psychol Aging 2014; 29(4): 891-906.
[http://dx.doi.org/10.1037/a0038244] [PMID: 25402337]

[46]   d'Orsi E, Xavier AJ, Steptoe A, *et al.* Internet use and physical activity can improve 10-year survival of older adults: Results from the English longitudinal study of aging. The 2015 Ageing Summit. 11 February 2015; London. 2015.

[47] Alzaid A, Alsulami M, Al-Maraghi A, Komal K. Examining the relationship between the internet and life expectancy. J Internet e-Business Studies 2014; 11

[48] Cook DJ, Augusto JC, Jakkula VR. Ambient intelligence: Technologies, applications, and opportunities. Pervasive Mobile Comput 2009; 5(4): 277-98.
[http://dx.doi.org/10.1016/j.pmcj.2009.04.001]

[49] Ramos C, Augusto JC, Shapiro D. Ambient Intelligence—the Next Step for Artificial Intelligence. IEEE Computer Society 2008; 23(20): 15-8.

[50] Riva G. Ambient intelligence in health care. Cyberpsychol Behav 2003; 6(3): 295-300.
[http://dx.doi.org/10.1089/109493103322011597] [PMID: 12855086]

[51] Afonso RA, dos Santos Brito K, Holanda do Nascimento C, Campos da Costa L, Álvaro A, Cardoso Garcia V. (Br-SCMM) Brazilian Smart City Maturity Model: A Perspective from the Health Domain. Stud Health Technol Inform 2015; 216: 983.
[PMID: 26262285]

[52] Kamel Boulos MN, Tsouros AD, Holopainen A. 'Social, innovative and smart cities are happy and resilient': insights from the WHO EURO 2014 International Healthy Cities Conference. Int J Health Geogr 2015; 14: 3.
[http://dx.doi.org/10.1186/1476-072X-14-3] [PMID: 25588543]

[53] Jiang S, Zhu X, Wang L. EPPS: Efficient and Privacy-Preserving Personal Health Information Sharing in Mobile Healthcare Social Networks. Sensors (Basel) 2015; 15(9): 22419-38.
[http://dx.doi.org/10.3390/s150922419] [PMID: 26404300]

[54] Available at: http://www.apho.org.uk/default.aspx?QN=P_HEALTH_PROFILES , [accessed 4 December 2015];

[55] Konstantinidis EI, Bamparopoulos G, Billis A, Bamidis PD. Internet of things for an age-friendly healthcare. Stud Health Technol Inform 2015; 210: 587-91.
[PMID: 25991216]

[56] van Hoof J, Kort HS, Rutten PG, Duijnstee MS. Ageing-in-place with the use of ambient intelligence technology: perspectives of older users. Int J Med Inform 2011; 80(5): 310-31.
[http://dx.doi.org/10.1016/j.ijmedinf.2011.02.010] [PMID: 21439898]

[57] Brossette SE, Sprague AP, Hardin JM, Waites KB, Jones WT, Moser SA. Association rules and data mining in hospital infection control and public health surveillance. J Am Med Inform Assoc 1998; 5(4): 373-81.
[http://dx.doi.org/10.1136/jamia.1998.0050373] [PMID: 9670134]

[58] Louchez A. The Internet of Things Is a Secular Transformation. IOT Journal 2015. Available at: http://www.iotjournal.com/articles/view?13003 (accessed 1 December 2015)

[59] Dallas E. Having a Heart Attack or Stroke? Your iPhone Knows. 2015. Available at: http://www.everydayhealth.com/heart-health/having-a-heart-attack-or-stroke-your-iphone-knows-6152.aspx (accessed 6 December 2015)

[60] Hussi T. Reconfiguring knowledge management: Combining intellectual capital, intangible assets and knowledge creation. ETLA Discussion Papers, The Research Institute of the Finnish Economy (ETLA) 2003; 849 Available at: http://www.econstor.eu/bitstream/10419/63966/1/361986475.pdf

[61]   Mehralian G, Nazari JA, Akhavan P, Rasekh HR. Exploring the relationship between the knowledge creation process and intellectual capital in the pharmaceutical industry. Learn Organ 2014; 21(4): 258-73.
[http://dx.doi.org/10.1108/TLO-07-2013-0032]

[62]   Mahmoudi R, Mortazavi SM, Safari S, Nikseresht M, Mozdarani H, Jafari M. Microwave Electromagnetic Radiations Emitted from Common Wi-Fi Routers Reduce Sperm Count and Motility. International Journal of Radiation Research 2015. Available at: https://www.researchgate.net/publication/277248261_Microwave_Electromagnetic_Radiations_Emitted_from_Common_Wi-Fi_Routers_Reduce_Sperm_Count_and_Motility (accessed 20 December 2015)

[63]   Mortazavi SM, Tavassoli A, Ranjbar F, Moammaiee P. Effects of laptop computers' electromagnetic field on sperm quality. J Reprod Infertil 2010; 11(4): 251-8.

[64]   Pianka ER. On r and K selection. Am Nat 1970; 104(940): 592-7.
[http://dx.doi.org/10.1086/282697]

[65]   Day JJ. New approaches to manipulating the epigenome. Dialogues Clin Neurosci 2014; 16(3): 345-57.
[PMID: 25364285]

[66]   Otani A, Kojima H, Guo C, Oishi A, Yoshimura N. Low-dose-rate, low-dose irradiation delays neurodegeneration in a model of retinitis pigmentosa. Am J Pathol 2012; 180(1): 328-36.
[http://dx.doi.org/10.1016/j.ajpath.2011.09.025] [PMID: 22074737]

[67]   Bruch H, Ghoshal S. A bias for action. Boston, USA: Harvard Business School Press 2004.

Appendix

PART 1. COGNITIVE AGEING QUESTIONNAIRE

Although not scientifically validated, this is a general-purpose questionnaire that can be used in the clinic or at home, aiming to evaluate certain aspects of cognition in ageing. The questions cover a wide spectrum of ordinary, everyday activities and reflect an element of hormetic stimulation and challenging cognitive behavior. The concept of *power law* (frequent mild activities, occasional moderate ones and rare intense ones) is also reflected. This can be used as a tool to inform the user about the value of cognitive challenges, and the role of hormesis in daily life. Part 2 of the Appendix is then about examples of activities one can participate in, so that to try and improve cognitive and physical functions, thus providing a practical guide to complement the questionnaire.

Cognitive Ageing Questionnaire

PLEASE TICK ONE

1. Do you watch (listen to) the news on TV (radio)?

1. regularly
2. frequently
3. occasionally
4. never

2. Do you believe that you will be able to develop new skills when you are much older?

1. yes
2. maybe
3. I don't know
4. no

3. Do you have to remember many tasks or events during your day?

1. all the time
2. frequently
3. occasionally
4. no

4. Do you get bored with ordinary routine?

1. yes, I like a change
2. sometimes
3. don't know/undecided
4. no, I love everyday routine

5. Are you satisfied with your job (or with not working)?

1. not at all
2. not sure
3. sometimes
4. yes, I am satisfied

6. Do you have to think creatively and use your brain during your day?

1. yes, all the time
2. frequently
3. occasionally
4. no, almost never

7. During the past two years how frequently have you been to intellectually stimulating courses (*e.g.* evening or weekend courses, part-time home study)?

1. regularly
2. frequently
3. sometimes
4. never

8. How frequently do you read science, art, philosophy, or other serious books (or online)?

1. every day
2. two or three times every week
3. two or three times every month
4. almost never

9. Do you find that your memory is getting worse as you get older?

1. definitely
2. somewhat
3. not sure

4. no

10. Do you believe that your intelligence is frequently letting you down?

1. yes
2. sometimes
3. not sure
4. no

11. Do you keep yourself informed of the latest developments in science, fashion, politics or world affairs?

1. yes, every day
2. frequently
3. sometimes
4. no, almost never

12. Do you daydream about positive life events (either past or future)?

1. regularly
2. frequently
3. sometimes
4. never

13. Are you happy with the way you are stimulated intellectually during your day?

1. yes
2. frequently
3. sometimes
4. no, too much or too little stimulated

14. In general, do you believe that people's memory seriously gets worse as they grow old?

1. yes
2. frequently
3. I don't know
4. no

15. a) Does intelligence worsen with age? Yes No

1. Does learning ability decline with age? Yes No

2. Do you believe that there is nothing you can do about memory loss? Yes No

3. Do you specifically avoid using your brain in order to avoid wear and tear? Yes No

16. Do you do crosswords, puzzles, memory exercises, brain training or using the internet to keep you cognitively stimulated?

1. yes, daily
2. once or twice every week
3. once or twice every month
4. no, almost never

17. Do you feel uneasy when having to handle new equipment (video recorder, computer, a new app *etc.*)?

1. yes
2. sometimes
3. not sure/no opinion
4. no, I look forward to the challenge

18. Are you unable to make decisions easily?

1. yes, I am
2. sometimes
3. not sure
4. no, on the contrary

19. Do you fall behind with your schedule at work or at home?

1. almost never
2. sometimes
3. many times
4. all the time

20. How much time do you set aside for you hobbies?

1. every day
2. about twice a week
3. about once or twice a month
4. none at all

21. Do you feel depressed, lonely or unhappy?

1. all the time
2. frequently
3. occasionally
4. almost never

22. a) Do you usually sleep well? Yes No

b) Is your appetite usually good? Yes No

c) Do you frequently feel like crying? Yes No

d) Have you given up hope? Yes No

Now check your score. Award yourself the points corresponding to your answers:

1. a)0 b)2 c)8 d)10
2. a)0 b)2 c)8 d)10
3. a)0 b)2 c)3 d)10
4. a)0 b)3 c)8 d)10
5. a)10 b)8 c)7 d)0
6. a)0 b)2 c)8 d)10
7. a)0 b)2 c)8 d)10
8. a)0 b)3 c)6 d)10
9. a)10 b)5 c)4 d)0
10. a)10 b)8 c)3 d)0
11. a)0 b)3 c)8 d)10
12. a)0 b)2 c)8 d)10
13. a)0 b)6 c)8 d)10
14. a)10 b)7 c)3 d)0
15. a) yes = 8 no = 0, b) yes = 8 no = 0, c) yes = 10 no = 0, d) yes = 10 no = 0
16. a)0 b)2 c)5 d)10
17. a)0 b)3 c)6 d)10
18. a)10 b)7 c)6 d)0
19. a)0 b)3 c)7 d)10
20. a)0 b)3 c)7 d)10
21. a)10 b)8 c)5 d)0
22. a) yes = 0 no = 5, b) yes = 0 no = 6, c) yes = 8 no = 0, d) yes = 10 no = 0

The lower the score, the more likely it is that your cognitive function and is healthy. You should make a record of your score and then retake the questionnaire at regular intervals, for example every six months. You can then compare your score with the original and see whether your mental attitude or cognition are improving. A persistently low score indicates that your brain and mental attitude are healthy.

COMMENTS ON THE ANSWERS

1. Keeping yourself informed about the national and international situation implies that you have a certain interest in the world around you which keeps the brain nicely stimulated. Do make a point of watching the news and then, if possible, discuss these with a neighbour or friend. Also spend some time online exploring new concepts, and sharing information which makes others act on it.

2. Given the right help and assistance, even very old people can learn computing and develop many other skills. Memory and mental faculties should be exercised but, as everything in life, this should be within reason, should not be excessive or prolonged. Rather, frequent light mental exercises could be mixed with less frequent moderate ones.

3. Day to day and hour to hour memory (short term memory) is important and needs to be exercised. If you are not using your memory frequently it may deteriorate later on in life. Try remembering some commonly used telephone numbers instead of writing them down.

4 and 5. Being satisfied with your everyday life is the basis of happiness. If you are not happy with your routine you now will also feel unhappy later on in life. However, this does not mean that you should sit back and enjoy what you have. You also need to seek new adventures, new situations and explore new ideas, in a way it makes you feel comfortable.

6. All faculties of our brain need to be exercised. Research suggests that a brain which is stimulated and challenged regularly will perform well even in very old age. Stimulating activities can be:

- thinking how to reposition the furniture in your sitting room
- writing a letter to your local paper on an issue
- learning how to use your local library, or a new computer

7. Studying keeps the brain active and disciplined. It also makes you feel better, particularly at the end when you receive the qualification. There are hundreds of courses for all abilities. Ask at your local library or adult education department, or explore hundreds of free courses online.

8. Reading books shows an interest in learning new things which in itself keeps the brain

active. It also means that in later life you will have something to do (*i.e.* read books), therefore boredom will be less likely. It is never too late to start reading stimulating books regularly.

9. Memory may worsen with age, more so if it is not exercised regularly. A certain degree of memory loss is normal. If you think that your memory is getting seriously worse you should see your doctor for evaluation.

10. If you do not believe in yourself then your chances of brain problems later in life will increase. If your brain is letting you down, why not exercise it? It may also be possible to consider oral supplements which may help you achieve a state of positive stress. These are called 'hormetins'. Ask a suitably qualified health practitioner for details.

11. By keeping yourself informed you maintain the brain active and stimulated. You will also get information about new anti-ageing treatments which you may find useful in maintaining your health.

12. It is wrong to think that day-dreaming about past events is damaging. People of any age need to think and review their lives constantly. This is particularly true in later life. Reminiscence therapy helps people put their lives in perspective and improves their current role in life. Day-dreaming about future events increases motivation and well-being. Also, meditation and mindfulness exercises are essential in improving several health aspects of your brain.

13. Intellectual stimulation during the day keeps your brain in good form. However, over-stimulation can cause too much strain, headaches and other unnecessary discomfort. When you are satisfied that you have had enough then it is time to stop for the day.

14. An older person may use different ways of dealing with a mental problem than a younger person but this is not abnormal. In some cases, Alzheimer's disease is the cause of mental impairment but this is a disease and not necessarily a normal ageing event. "Senility" is not an inevitable part of ageing.

15. The brain does not slow down appreciably but it uses different techniques to deal with problems. Properly performed IQ studies show that intelligence does not decline significantly in healthy elderly people. With proper training healthy elderly people may be able to maintain good cognitive functions for many years.

16. If you use your brain by doing activities like chess, puzzles, writing *etc.*, you may reduce the chances of brain failure in later life. Brain stimulating activities need not be very hard, an easy crossword will do. Activities which put a lot of pressure on the brain may cause

unnecessary strain. However, always remember that, for best results, you need to perform frequent light mental activities, some moderate ones and rarely, some heavy ones.

17. It is often difficult to handle technical equipment, and unfamiliarity makes matters worse. However, you should try and keep up with technological advances which make life easier. The challenge of dealing with a new situation is a great way to stimulate your brain.

18. Taking a quick decision shows a sharp mind willing to take some risk. People who manage to carry these qualities with them into older age are fortunate. Those who don't, should exercise them. If however, you have been a slow decision maker throughout life, it will be very difficult to change.

19. The way you work reflects many aspects of your intellect and personality. Falling behind with everyday routine might mean a less organised approach to life which can cause trouble and confusion later on. Try to work in a way that looks chaotic and irregular but has underlying order and intent. This means you deliberately do irregular activities in order to achieve your aim, but this irregularity is planned and controlled. You should not act in a disorganized way but do well-controlled and varied activities.

20. Setting aside time for your hobbies is one of the most important steps to take in order to avoid stress. It also indicates that you have your priorities right, you care about yourself and therefore you keep yourself well.

21 and 22. These questions test for depression, which can present as memory loss. Treatment for depression is not always difficult and the results can be spectacular. If you think you are depressed see your doctor for treatment.

23. Physical exercise and a low cholesterol diet may improve blood flow to the brain. A family history of atherosclerosis (thickening of the arteries) may indicate an increased risk of brain failure. There are several nutrients and compounds which may help, if taken with medical supervision.

24. Recent research shows that low education is associated with an increased risk of Alzheimer's dementia. It is never too late to exercise the brain and try to prevent any further deterioration.

PART 2. EXAMPLES OF PRACTICAL COGNITIVE AND HORMETIC CHALLENGES

My colleagues and I have been using these exercises and advice for the general public for several decades. The intention is to place some aspects of the discussion from this book into a practical, simple framework and make it easy for people to use in their everyday life. The concepts of hormesis, environmental enrichment, power law, and positive challenges are incorporated in all of these exercises. One possible way these mental, social and spiritual challenges exert their hormetic benefits could be *via* the up-regulation of nerve growth factors and receptor-ligand interactions, increasing responsiveness of neural stem cells, and bolstering other neuroprotective elements such as reducing neuronal apoptosis or down-regulating the expression of amyloid alpha and synuclein. Another theoretically possible mode of action could be the induction of ultra-low secretion of glutamate or other neurotoxins/neurotransmitters, which have a protective hormetic effect on the brain. Population studies have confirmed that a combination of external stimulation such as activities of daily living (ADL), attendance at religious events, physical exercise and solitary leisure activities are associated with a reduced mortality in old age, whereas greater social networks and social engagements are associated with a reduced cognitive decline in elderly individuals.

Advice for the Public

Below are some examples of easy-to-perform practical cognitive and other exercises which can help improve the flow of information in your brain and achieve a state of positive stress. If you don't find an exercise stimulating enough (or if it is not 'stressful') then try to increase the difficulty by adding more complicated and challenging activities. If you feel pleasantly stimulated after the exercise then it means that you have achieved a suitable state of positive stress. Also if you do that particular activity every day anyway, then try to do something different that you don't do every day. The activity has to be varied and challenging for you. It should not create prolonged feelings of discomfort. Any type of stimulation, such as physical, mental, social, sensual and spiritual should be used, in order to keep your body and mind agile and energetic.

HORMETIC-TYPE CHALLENGES

NB. The order of how these exercises are presented has deliberately been diversified. The grouping together of similar activities (all nutritional together, all social together *etc.*) is, in itself, not hormetic!

Examples of physical stimulation

Day 1: Do 30 minutes tai chi in the morning and 20 minutes ballroom dancing in the evening

Day 2. Go for a 20 minute brisk walk, and then do 15 minutes yoga

Day 3. Play football or other ball games for 30 minutes in the afternoon

Day 4. No exercise, just follow everyday routine
Day 5. Try a new sport such as horse-riding, fencing or rowing
Day 6. To the park for 20-25 minutes of tree/stone or other weight lifting
Day 7. Aerobics for 20 minutes, followed by a new/unusual exercise (chi kung, twerking, lifting logs)
Day 8. No exercise

Varied socialisation. During a seven day period, you may want to consider the following plan

1st day: volunteer at your local charity or community centre
2nd day: litter picking with local scouts group
3rd day: no social activities, perhaps don't talk to, or interact with, anybody
4rd day: work with your local or national political party
5th day: online work, explore the internet, chat to friends through social media
6th day: physically visit relatives or friends. Try to make a new friend
7th day: light work with social media

Examples of unusual physical activities:

1. Try painting or drawing, even if you have never done it before (particularly then!)
2. On certain days or times, walk around your house completely naked (if children *etc.* allow it)
3. Choose a controversial subject and argue the side opposite to your own opinion
4. Listen to strange or weird music, or music you don't usually listen to
5. Tune into a foreign language radio station and try to guess the general meaning of the programme
6. For about 5-10 minutes, read a magazine holding it upside down. To make the exercise more difficult, hold it upside down in front of a mirror and read from the mirror image
7. Lie on the floor for 10 minutes and examine your surroundings from this unusual view point
8. Blindfold yourself and try to continue your usual activities in the house (if safe)
9. Read a magazine or newspaper that you do not usually read. Check out a weird website

More difficult exercises. Think about, and devise, exercises based on the concept of 'doing two/three demanding things at once'

For example: write down simple arithmetic (*e.g.* Addition: 2+2=4, 2+4= *etc.*) while someone is reading a story to you (then, remember and discuss the story)

Or: listen to music while surfing the internet AND using a grip exerciser on the other hand

Or: play table tennis while listening to the news AND arguing the side opposite to your own opinion

Or: use your non-dominant hand to write a good paragraph, while blindfolded (and chatting about something else)

Eat according to the season. We are used to eating the same foodstuffs regularly. But our innate biological patterns obey seasonal rhythms which we need to rediscover. Try to explore new and unusual foodstuffs and choose products which are in season, according to the month of the year: Blueberries, fish, tomatoes, sage and fennel for the summer, *etc*.

This introduces an element of irregularity and complexity in our food patterns that mimics nature. Also try to alternate your meals and include unusual items such as exotic fruit, game, or nuts.

Clock-less days. Rely on your natural timetable rather than clocks or watches. Set aside a day (maybe a Sunday) and prepare from the night before. Hide all watches, clocks *etc*. Wake up when you wake up, eat when you are hungry, and sleep when you are tired. This makes your body and brain able to deal with new situations and challenges, whilst trying to guess the natural time for you.

Technology-free days. You can try this activity on rare occasions, perhaps every six weeks.

Imagine that you live in the 16th century when electricity or gas have not been discovered. Try to live your life without using **any** modern facilities whatsoever.

EXAMPLE:

Don't use electric razor or toothbrush

Don't use electricity/gas to make cup of coffee or cook breakfast/lunch

Don't use telephone, TV, radio, computer *etc*.

Don't use anything that was not present in the 16th century

Instead:

Read a book, write a letter, chat to other people

Do some gardening, go fishing or go for a swim

Use your imagination to discover new games (with dice, cards or paper)

This exercise will make you think hard to identify items or services which you normally take for granted. It will then force your brain to develop alternative strategies for coping. The brain will be constantly on the lookout for any actions which are prohibited, and therefore it will keep continuously stimulated.

Stop the exercise when you feel that it is becoming quite uncomfortable to continue.

<u>Sense imagery</u> To help consolidate memory information already stored in your brain the technique of 'sense imagery' is used.

* Recall a face that you know really well. Then spend a few minutes thinking about the particular characteristics of that face and try to recreate the face in your mind. Instead of using a face you may think of:

1. A well-known scenery such as your house, your garden or your neighbourhood
2. Geometrical shapes like triangles, spheres, pyramids, stars
3. Tones of voices such as the voice of a well-known person. On this occasion you should really make an effort to feel the tone. Other hearing mental images are: a familiar voice calling you by your name, the shouts of the traders in the market, the noise of an airplane.
   - Try to recall in your mind the smell of some common objects such as your soap, your house, and others such as freshly cut grass, the smell of cooked eggs and bacon, surgical spirit *etc*.
   - Tastes. Can you recall, in turn, the taste of: bread, ketchup, salt, sugar, beer, vinegar?

Examples of social and spiritual stimulation

- Don't accept easy answers or explanations. Always look for the real reason behind each event or situation
- Strengthen your social bonds, friendships and family ties
- Take an active part in your society, by volunteering or by doing charity work
- Explore other religions
- Explore aspects of your own religion which you have not considered in the past
- Learn new methods of mediation
- Always look for new ideas, and then try to implement them

<u>A cognitive exercise.</u> Use your non-dominant hand to draw one of the following, upside-down.

A smiley face

A sad face

The numbers 5 and 7

Then: a clock (remember, upside down!). When you finish, turn it round. It should have the numbers in the correct sequence, evenly-spaced and correctly drawn. Try it again until you get it right.

This is a good example of beneficial brain stress. You make your brain work at different levels. First, by using your non-dominant hand (new neural pathways are stimulated). Second, by drawing upside down (your brain processes unusual visual inputs). Third, by trying to get the drawing evenly balanced (developing new judgment capabilities).

<u>An example of nutritional stimulation.</u> Apart from a calorie restricted regime which is not currently recommended in everyday situations, another nutritional stimulation is this:

If you are taking vitamins, herbals or nutritional supplements it may be better to try and take these at irregular intervals instead of taking the same dose every day at a regular time.

For instance, on one day you could take your vitamin E in the morning, your vitamin C before lunch, your co-enzyme Q10 tablets in the afternoon and your glucosamine in the evening. Next day take only two of the above together and one at night, missing one out. The following day do not take anything. The day after take one of the above at lunchtime and the three others in the afternoon. And so on. You should only do this under expert supervision.

CONCLUSION

By following a carefully planned but irregular and apparently haphazard lifestyle, you increase the amount of meaningful stimulation to your body and mind, therefore lending a hand in maintaining the complex biological reserves of your body. This may help you ward off ageing and disability. Although, to an untrained observer, may appear chaotic, it should in fact be planned very carefully with logic behind each step.

Using a combination of suitable medication with lifestyle measures can go a long way in helping you defeat some signs of ageing. What matters is you. If you feel satisfied with a particular nutrient or drug, if you enjoy your lifestyle and if you are content with your plan of action, then so be it. You only have a short time left on this earth and you should make the most of it.

But also remember that you can influence that time, up to a certain point. Choose wisely, and you will increase it both in quality and in quantity. Choose poorly, and you will make it even shorter.

SUBJECT INDEX

A

Acid 44, 53, 54
   arachidonic 54
   ferulic 44
   free fatty (FFA) 53
Actions 38, 50, 51, 52, 78, 207, 208
   physical 38, 50, 52, 78, 207, 208
   pro-ageing 51
Action video games (AVGs) 73
Activities 59, 215
   anti-ageing 59
   information-rich internet-based 215
Adaptation 60, 62, 99, 107, 112, 159
   environment-induced 107
   mechanisms 60, 62, 99, 112, 159
Ageing 77, 83, 89, 101, 109, 112, 129,
   142, 185, 204, 217, 221, 227
   management 217
   mechanisms 101
     human 89
   -modifying factors 204
   nodes 185
   parameters 83
   parents 129
   process 77, 112, 142, 221, 227
   stops 109
   system 185
Agents 3, 4, 8, 9, 19, 38, 43, 47, 49, 50,
   51, 54, 55, 58, 77, 105, 132, 133,
   143, 146, 147, 163, 171, 173, 174,
   175, 177, 178, 181, 182, 183, 184,
   191, 209, 213, 215
   human 209
   interacting 177, 178
   xenohormetic 58

Age-related 7, 8, 15, 23, 28, 43, 44, 46,
   47, 55, 63, 87, 88, 89, 101, 103, 104,
   122, 125, 128, 129, 138, 142, 143,
   164, 201, 202, 203, 206, 210, 211,
   220, 221
   damage 15, 43, 44, 103, 122, 125, 201
   degeneration 8, 23, 28, 47, 55, 63, 87,
     88, 89, 101, 129, 138, 142, 143, 164,
     206, 210, 211, 220, 221
   dysfunction 7, 46, 89, 104, 128, 202,
     203
Alzheimer's dementia 114
Ambient Intelligence (AmI) 159, 163,
   201, 204, 209, 213, 214
Aminoguanidine 48, 49
Amygdala 27, 80
Anabolic steroids 53
Anandamide 55, 56
Anterior insula 23, 24, 27, 80
Anti-ageing 48
   actions, potential 48
   pills 48
Anti-stress gene 11
Apoptosis, germ cell 127, 133

B

Benefits 73, 164, 206, 207
   cognitive 73, 164
   physical 206, 207
Bioflavonoids 45
Biologically-usable information 113
Blood-brain barrier 111
Body, badly-functioning 72
Boltzmann's constant 195
Borivilianum 56

* 9 7 8 1 6 8 1 0 8 3 3 6 0 *